China's Big Science Facilities

"Big science" facilities are major elements of science and technology infrastructure, and important symbols of China's scientific and technological development. This popular science book series presents the background, history and achievements of the Chinese Academy of Sciences in terms of constructing and operating big scientific facilities over the past few decades.

The series highlights the major scientific facilities constructed in China for pioneering research in science and technology, and uses straightforward language to describe the facilities, e.g. the fully superconducting Tokamak fusion test device (EAST), the National Protein Science Research Facility, Lanzhou Heavy Ion Accelerator, Five-hundred-meter Aperture Spherical Telescope (FAST), etc. It addresses the respective facilities' research fields, scientific backgrounds, technological achievements, and strategic and fundamental contributions to science, while also discussing how they will improve the development of the national economy. Supplementing the main text with a wealth of images and linked videos, the book offers extensive information for members of the general public who are interested in scientific facilities and related technologies.

More information about this series at http://www.springer.com/series/16530

Genming Jin · Guoqing Xiao

Editors

Probing into the World of Nuclei

Heavy Ion Research Facility in Lanzhou

ZHEJIANG EDUCATION PUBLISHING HOUSE

Springer

Editors
Genming Jin
Institute of Modern Physics
Chinese Academy of Sciences
Lanzhou, Gansu, China

Guoqing Xiao
Institute of Modern Physics
Chinese Academy of Sciences
Lanzhou, Gansu, China

Translated by
Mao Li
Beijing Academy of Social Sciences
Beijing, China

Yaxian Wang
Beijing Academy of Social Sciences
Beijing, China

ISSN 2662-768X   ISSN 2662-7698 (electronic)
China's Big Science Facilities
ISBN 978-981-16-0717-2   ISBN 978-981-16-0715-8 (eBook)
https://doi.org/10.1007/978-981-16-0715-8

Series Foreword

As a new round of technological revolution is bourgeoning, it will exert a direct impact on the survival of a country whether or not it can gain insight into the future technological trends and grasp new opportunities from the revolution. In the face of the major opportunities in the twenty-first century, China is intensively formulating the innovation-driven development strategy and building an innovation-based country in this critical era to achieve a moderately prosperous society in an all-round way.

Scientific and technological innovation and popularization remain two wings for innovation-driven development of a nation. In particular, popular science affects the awareness of the general public for science and technology as well as social and economic development. Scientific education is thus highly practical for implementing the innovation-driven strategy. Contemporary science pays more attention to public experience and engagement. The word "public" covers various social groups that exclude those in scientific research institutions and departments. The "public" also includes decision-makers and management personnel in government agencies and enterprises, media workers, entrepreneurs, science and technology adopters, etc. Barriers that impede the innovation-driven strategy will emerge if any group falls behind this new revolution; avoiding and removing the possible barriers will strategically improve the quality of human resources, enhance mass entrepreneurship and innovation and build a moderately prosperous society in an all-round way.

Science workers are primary creators of scientific knowledge who undertake the mission and responsibility for science popularization. As a national strategic power in science and technology, Chinese Academy of Sciences (CAS) has always attached equal importance to this mission in addition to scientific innovation and incorporated the mission into key measures of the "Pioneering Action" Plan. CAS enjoys rich and high-end technological resources, such as the high-caliber experts represented by CAS members, advanced research facilities and achievements represented by the Big Science Project, excellent scientific popularization base represented by the national scientific research and popularization base. With these resources in place, CAS implements the "High-level Scientific Resource Popularization" Plan to transform the resources into popular facilities, products and talents to benefit trillions of the public. Meanwhile, CAS launches the "Science and China" program, a scientific education plan, to mobilize more effectively the "popularized high-end scientific research resources" for scientific education targeted at the public and the integration of science and education.

Scientific education requires not only the dissemination of scientific knowledge, approaches and spirit to enhance overall scientific literacy of the country but also the creation of scientific environment to enable scientific innovation to lead sustainable and sound social development. For this reason, CAS cooperates with Zhejiang Education Publishing House to launch the CAS Scientific and Cultural Project. This is a large-scale scientific and cultural communication project on the basis of CAS research findings and expert teams to improve the scientific and cultural quality of the Chinese citizen in an all-round manner and to serve for the national strategy of rejuvenation by advancing science and technology. On the basis of the target group, the project is categorized into two series, i.e. the Adolescent Scientific Education and the Public Scientific Awareness, respectively, for the adolescent and the general public.

The Adolescent Scientific Education series aims to create a series of publications that draw on latest scientific research findings and introduce the status quo of scientific development in China, to cultivate the adolescents' interest in science study, to educate them about basic scientific research approaches and to inspire them to develop rational scientific way of thinking.

The Public Scientific Awareness series aims to educate the general public about basic scientific approaches and the social significance of science and encourage the public to engage in scientific affairs, thus the project will enhance the capacity the public of conscientiously applying science to their life and production activities, improve efficiency and promote social harmony. In the near future, publication series of CAS Scientific and Cultural Project will constantly come out. I hope that these publications will be welcomed by the reader and that through coordination among CAS science workers, science icons such as Qian Xuesen, Hua Luogeng, Chen Jingrun and Jiang Zhuying will be more familiar to the public. As a result, the truth-pursuing spirit, rational thinking and scientific ethics will be fully promoted, and the spirit of science workers in courageous exploration and innovation stands eternally in the history of human civilization.

July 2016

Chunli Bai
President of Chinese Academy of Sciences
Secretary of Leading Party Members' Group

Preface

My dear friends, do you know what is the smallest particle that makes up all substances in the world? You may say it is the atom. Then what is the atom made up of? What is the substance that constitutes an atom? What does it look like? How can we use the atom? If you are not sure of the answers, just check this book.

An atom is like a fort. Its wall consists of electrons, although not that compact and strong. Each castle only has one inhabitant—a dwarf (nucleus). In the world of dwarfs, there are nearly 3,500 different inhabitants. They draw their own map in the blue ocean. Like our world, their world also has many rules and disciplines. Every dwarf voluntarily abides by the rule with which an orderly world is built. Yet in this world, every dwarf has its own characteristics and will try to play their roles.

To know the origin of the dwarf, create more dwarfs and make them to do things for us, we must first of all mobilize them and make them a fighting force. To this end, scientists have created various dwarf force training bases—accelerators. From this book, you can know the internal situation of these bases and how to train the dwarf force.

Where do the dwarfs come from? A few were born in the initial chaotic state of the universe. More were produced during the evolution process of the stars from birth to decease. However, there are not many—only 287 kinds of dwarfs—left for us to use and research. Over 3,000 of them have been created and carefully studied by scientists in a laboratory over the past seven or eight decades. The dwarf force can help us to kill tumors, improve crops and flowers, Chinese herbal medicine, as well as microorganisms so that they are better used for humanity. The dwarf force can also provide important assistance for the safety of artificial earth satellites, the moon-lander "Chang'e", explorers to Mars, and astronauts!

For more details, please check and read this book. If you have any opinion and advice, please inform us and we will revise it based on your feedback.

Come and visit the Institute of Modern Physics (IMP) of the Chinese Academy of Sciences—a renowned nuclear physics institute in China—if you want to know more stories about dwarfs, training bases and training process of the dwarf force, and achievements of the dwarf force.

Lanzhou, China
July 2017

Genming Jin

Contents

Visiting the Dwarfs in the Atomic Castles

Genming Jin and Guoqing Xiao

Go to the atomic castles and look for their owners. Learn more about the castle owners and their family members. Watch the dwarfs' sports games and their acrobatic performance.

1 Visiting the Dwarfs in the Atomic Castles

(1) Flying Through the Walls of the Atomic Castles

Nowadays, many people travel around during their holidays. On weekends, they may drive somewhere near their home to have fun; during long vacations, they may travel long distances on highways, airplanes or trains, to name just a few. Some like walking on a beach, enjoying gentle breezes and the charming blue sea. They may also jump into the sea for a joyful swim. Some prefer mountains for breathtaking landscapes. Have you, then, ever dreamed of turning yourself into a tiny little person, boarding a latest micro helicopter to travel around the microscopic universe for the fantastic landscapes? If you have such a wish and the gut, just go ahead! (Fig. 1).

Turn smaller! A fragrant breeze reached me, and I fell asleep. I did not know how long it took before I heard a gentle voice calling: "Open your eyes, we are arriving at the airport!" I rubbed my eyes and looked around to see who was talking, only to find mountains floating in the air. It seemed like I had landed on the planet of avatars. Though covered with no trees or grass, the mountains were all shining with bright light. I took a closer look at the ground below and found rolling hills with cracks. Beside me landed a helicopter, whose propeller was spinning around. "Maybe it's waiting for me to get on board", I thought (Fig. 2).

G. Jin (✉) · G. Xiao
Institute of Modern Physics, Chinese Academy of Sciences, Lanzhou, Gansu, China
e-mail: jingm@impcas.ac.cn

© Zhejiang Education Publishing House 2021
G. Jin and G. Xiao (eds.), *Probing into the World of Nuclei*, China's Big Science Facilities, https://doi.org/10.1007/978-981-16-0715-8_1

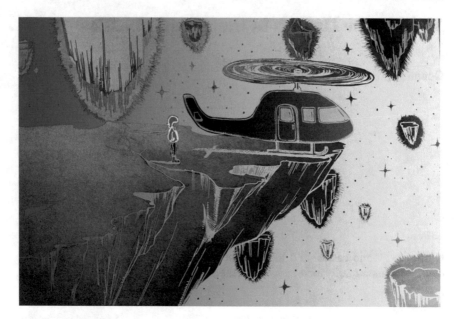

Fig. 1 Atomic world

Fig. 2 The electron is everywhere

The helicopter took off the moment I got on board. Then a uniquely sweet female voice came from this driverless helicopter: "Hello, little lady! Welcome on board our helicopter to the microscopic universe. We will introduce to you the sceneries along the flight. First, I'd like to tell you that our helicopter is not too small compared with the electron." Oh, man! A helicopter was compared in size to an electron! But what on earth was this interesting electron?

"You must know electricity, which is crucial in modern society. The electricity used in daily life is the electric current flowing through the wires. In 1897, a British named Thomson discovered the electron during an experiment. The term 'electron' was coined yet earlier by another British named George Stoney. The electron is quite small, just like a grain of sand compared to our helicopter. It has mass of around $1/10^{30}$ kg (9.11×10^{-31} kg). Small as it is, the electron is an essential part of the atomic castle." As she spoke, I heard a thump. "You might have heard a noise, it's the sound of a flying electron crushing into the porthole of the helicopter."

While listening to her, I looked through the porthole. Mountains were floating, and castles of all sizes were flying around rapidly. Oddly enough, some castles were alone, some bounded in three, but most in pairs, snuggling up together like lovers. The helicopter was dodging to left and right, for fear of crushing into some intimate lovers.

I was trying to figure out why all this happened. Just then, the guide explained in a pleasant voice as if she was able to read my mind: "You may have noticed that there are castle-like round buildings as well as the floating mountains out there. These buildings are gaseous molecules in the air, and oxygen is one of them. The English letter 'O' represents the atom of oxygen. Each atom of oxygen will find a companion either of its same or a distinct type. Only in this way can they form a solid castle. Look, the blue twin-balls are their castles, accounting for 21% of the total floating castles. Oxygen is an indispensable element for all things to grow (Fig. 3)". As I examined them, the blue pairs were constantly floating and spinning, as if performing a pas de deux. How beautiful their dance was!

"Nitrogen atoms are also a main component of the air. They, too, live in pairs. Those black twin-castles, similar in size to the blue ones, account for around 78% of the air. They are crucial elements in our daily life. For example, meat and eggs we eat everyday are rich in nitrogen. Yet nitrogen must bond to oxygen, carbon and other elements in a certain proportion to be edible." Oh, these dark dancers were equally charming. There were also some petite dancers, dressed in red, dancing swiftly through the black and blue couples. Though dancing alone, these graceful dancers were enjoying themselves.

"Well, who are those petite dancers in red?"

"Helium atoms. They are celibates."

The guide continued: "Look, two ladies in blue are dancing with a gentleman in silver grey. Those are carbon dioxide molecules. While small in number, they have a significant role to play. They can reflect heat back to the earth, contributing to a stable temperature of our planet. Too much carbon dioxide, however, will trap heat in the earth, leading to rising temperatures and various natural disasters." No wonder that

Fig. 3 The wondrous atomic world

in recent years, countries around the world have gathered together to sign agreements on cutting emissions of carbon dioxide (FIg. 4).

I saw castles one after another. Between the castles also floats freely small amounts of sand, never bonding together, but running into the castles occasionally (Fig. 5).

"The sand is the electricity, discovered 120 years ago by a man named Thomson. They are like-charged. We know, like-charged objects repel, and opposite-charged objects attract."

At that time, a castle rushed by our helicopter. I hurriedly filmed this incredible scene. As I was watching the video replay, I found the castle quite strange. It looked like a giant ball with thick "walls"; but a closer look revealed that the walls seemed to be heavy clouds. What were these "walls" made of and how were they built?

"'Nothing ventured, nothing gained.' Let's fly for the castle", I said. Yet however hard we tried, we still could not catch up with the fast-moving castles. What a pity!

"Don't worry, we can go to the Steel World where rows of castles are moving at a moderate speed." The guide's words reassured me.

The massive mountain suddenly disappeared as our helicopter approached it. We were plunged into darkness, down, down and down. At that time, the helicopter's high-powered searchlight was turned on. With the light, I saw a striking sight: rows of shining castles were floating in an open and transparent space.

"How could castles possibly move?" I asked in perplexity.

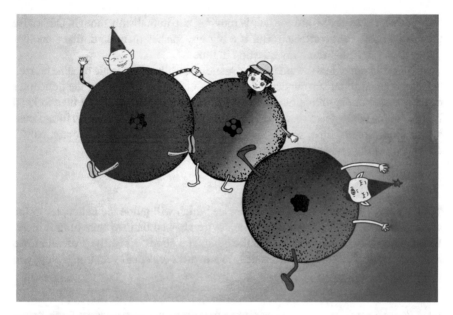

Fig. 4 Atomic castles in carbon dioxide molecules

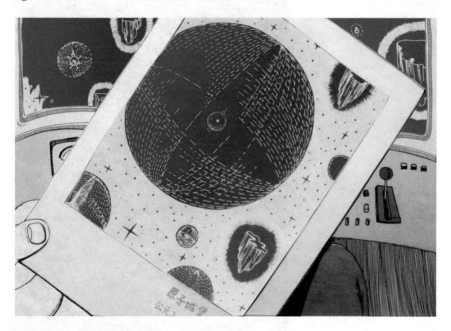

Fig. 5 The atomic castle (atom)

"Indeed. These castles are constantly moving, not dramatically though. Otherwise their neighbors will be affected. But if a big guy pushes them hard, that's another thing." The guide answered my question in time.

As the helicopter flew toward another castle, the walls of the castle became blurred and we seemed to be surrounded by a mist of electrons. Flying on, the helicopter came across more speeding electrons. It had to dodge every now and then to keep flying. After a while, there seemed to be less electrons and I was shrouded in darkness again. From time to time, some shining electrons passed by the helicopter. Now I could see nothing; where was helicopter heading for?

(2)   Visiting Dwarfs in the Atomic Castles

"Don't worry. We have a positioning device which will guide the helicopter to the center of the castles." After a while, the device showed that we were almost there. But still, we saw nothing. Suddenly, the helicopter shook violently. "Attention, we have arrived at the center of the castles. The owners are small yet strong. Now you could talk to them freely (Fig. 6)."

Fig. 6 The dwarf (the nucleus)

I looked out, only to find something like basketball in the distance. Slowly, we approached the "basketball".

Oh! It turned out to be a strange-looking man with a fat belly, a tiny head and short and thin arms.

I got out of the helicopter and greeted him politely, "hi, sir!".

"Hello! I must tell you that I'm highly charged, you'd better stay away from me. But if you are wearing an insulating suit, a handshake is welcome."

"OK! May I ask how many people live here? Why I didn't see anybody else?"

"Well, I'm the only person living here, strange, huh? But here, one castle only hosts one person."

"How big is your castle? And what's your name, please?"

"The place where I live is called iron atomic castle. It's huge. Here's an analogy. If I was magnified to one meter and so was my castle in the same proportion, the inner wall would be a few kilometers away from me. The thickness of walls varies in every castle, and mine are around several dozen meters thick. You may have noticed before you entered here that the castles are arranged in order (Fig. 7). Every castle has thick walls made of several layers of floating electrons. As I live at the heart of the atomic castle, I'm named atomic nucleus. I'm so small compared to the castle, so you can call me dwarf!"

"Did our helicopter hit you? Are you hurt?"

"Ah, I was wondering what rubbed my belly just now."

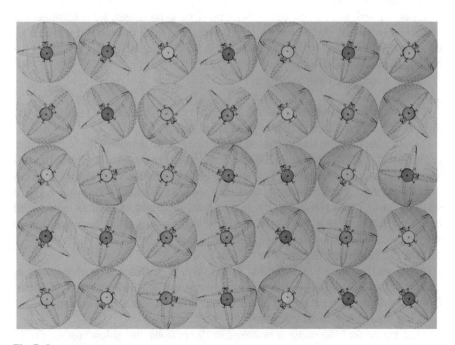

Fig. 7 Iron atom array

"Sorry! But you are so tiny, how come our helicopter shook violently and nearly crushed when it hit you?"

"Small as I am, I'm very strong and heavy. Although the wall is thick, it only has four thousandth the mass of mine. If I grew as big as you guys, I would have a mass of three hundredth that of the Earth." (The Earth's mass is around 60^{24} kg and its density 5500 kg/m^3; the nuclear density of iron around 2.8×10^{22} kg/m^3.)

"Why is your belly so big? Do you eat good things every day?"

"My belly is always this big, but I've got concrete things in it. There are two types of small balls of the same size, one with positive charge, the other without charge. The small balls positively charged are protons which total 26. It is because of them that I'm charged. The other uncharged balls are neutrons, 30 in total. Although they have no charge, I cannot live without them."

"Thank you! I've learned quite a lot from you. Goodbye!"

We went back into the helicopter and flew on.

(3)   The Dwarfs' Brothers

Flying through the fast-moving electron walls, we left this castle and came to another one. Again, through the electron walls, we struggled to finally find the castle owner. Just like the dwarf we had met, this castle owner had 30 neutrons and 26 protons in his belly. We then visited several dozen castle owners, most of whom were brothers from a multiple birth.

Later, we arrived at a new castle. With previous experiences, we soon found the castle owner. At first glance, he was no different from other dwarfs. But when I took a closer look, I noticed something different in him. To figure it out, I got out of the helicopter and talked to him.

"Excuse me, may I call you dwarf?"

"Sure!"

I've just visited some castle owners who look rather like you. But I still think you guys are a little bit different. Could you tell me why?"

"You are lucky enough to see us—the minorities. Most dwarfs living in the castles are identical, with 26 protons and 30 neutrons in their belly, like multiples. But I'm somewhat different from them. You see, my belly is smaller. That's because I have only 28 neutrons, 2 neutrons fewer than they have. So I'm their little brother. I have an elder brother who has one more neutron than those buddies you've met. We all have the same number of protons in our belly, and we occupy the same position on the periodic table, so we are siblings (Figs. 8, 9 and 10)."

"Do you have other brothers?"

"Yes, but they are so naughty that they always play all kinds of tricks, such as spitting out an electron or swallowing one from the walls. This is no laughing matter. Once they do so, they are no longer our brothers, but members of other families."

"How many brothers do you have?"

"Let me think. My youngest brother has 19 neutrons, my eldest brother 47. So altogether I have 28 brothers, of whom four are quiet and the rest are quite playful. Those with the most or fewest neutrons can play tricks in a fastest way."

Fig. 8 The dwarf and his brothers (isotopes)

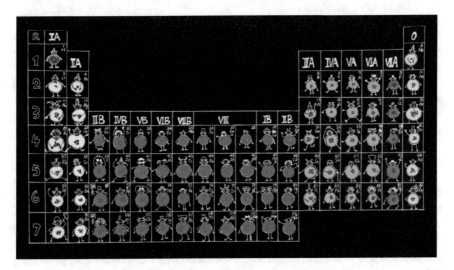

Fig. 9 The family of the dwarfs (the periodic table)

These dwarfs were so naughty!

"Thank you for telling me so much. Could we visit your other brothers?"

"Sure. Yet I don't know which castles belong to them. You have to look for them on your own. Wish you good luck! Remember, the castles of our families are of the

Fig. 10 The periodic table compiled by Mendeleev

same size. Don't look for them in the small or big castles, of which the owners are not our family members!"

"Thank you for reminding us. Bye!"

We bid farewell to the dwarf and flew to other castles. There came the sweet voice again.

"The castle owner mentioned his brothers. Now I'd like to tell you more about these brothers. But before that, let me first introduce the periodic table."

A picture appeared on the television screen. There was a table with an element's name and some numbers in each cell.

"Look, this is the periodic table. Just like the seating arrangement of the heroes in the Chinese novel *All Men Are Brothers*, each element takes a seat which represents its position in the table.

Their mass and properties are all written on the seats. From left to right in the rows, changes of the properties are similar. Take the second row for example, the leftmost element is lithium, a highly reactive metal which can be used for producing lithium batteries. On the right side of lithium is beryllium. As metallic elements, these two elements are placed together. Next to beryllium is boron, and next to boron carbon. Boron and carbon are both metalloid elements. On the right side of carbon are nitrogen, oxygen and fluorine in sequence, which are all gases. The rightmost element is neon, a chemically inert gas. This element was injected into small glass capsules with electrodes to make neon lamps in the past.

   With lamps that are more energy efficient, people rarely use neon lamps today. From left to right in the rows, the order of elements is: metals, nonmetals and gaseous elements. Noble gases, which never combine with other elements, are to the right-most. Each row represents a period, so this table is called periodic table which reflects periodic changes of the elements' properties. The table has 118 elements, outnumbering the heroes in *All Men Are Brothers* by 10. Among these elements, only 90 can be found in the Earth's crust; the rest 27 have been synthesized in laboratories during the past 70 years."

   After a pause, the guide continued.

   "Do you know the origin of this table? As far back as 1869, 63 elements and their respective mass and properties had been discovered. Back then, a Russian chemist professor named Dmitri Ivanovich Mendeleev, after considering for a long time, decided to arrange these elements based on their mass and properties. He found that the properties of these elements showed a periodicity as their mass increased. So, he grouped them according to the recurring patterns. Elements in each group were listed in a column in ascending order of mass. The columns were then arranged from left to right, based on the mass of the first element in each column. The same is true of the military parades, where soldiers assemble in columns and march in formations. That's how Mendeleev compiled the world's first periodic table. To honor his contribution, the table is also called Mendeleev's periodic table. As more and more elements were discovered, the periodic table has become what you see on the television."

   The helicopter flew over many castles whose owners had either 30 or 28 neutrons. But where was the brother somewhat different from these castle owners?

   "Don't worry, we'll find him." There came the pleasant voice again. OK, then, keep visiting. I would not give up.

   We entered another castle and managed to find its owner.

   "Hello", I greeted him in the distance.

   "Hello! What can I do for you?" replied the dwarf politely.

   "I'd like to visit the dwarfs' brother who is not born at the same birth."

Knowledge Link

Periodic Table

   The periodic table consists of 7 periods and 16 groups. Table rows are called periods and columns are groups (except for group VIII, which is arranged in three columns). Of the 7 periods, the 1st, 2nd and 3rd periods are long periods, the 4th, 5th and 6th are short periods and the 7th is incomplete period. There are 7 main groups (group IA, IIA, IIIA, IVA, VA, VIA and VIIA), 7 subgroups (group IB, IIB, IIIB, IVB, VB, VIB and VIIB), one group VIII and one group 0.

   The position of elements in the periodic table reflects not only the atomic structure but also the connection between the recurring patterns of chemical properties and elements. This complete system marks an important milestone in the advances of chemistry.

"You've come to the right place. I'm the one you are looking for."

"That's great! I've visited dozens of dwarfs, but they are all multiples. It's been so hard to find you!" I was so excited that my eyes become watery.

"Indeed. Among us, 91 out of 100 are multiples. I look quite like my brothers. So you won't find the difference between us if you don't take a close look. Now let's see if you could tell how I'm different from them."

"OK, I'll have a try." It took me about two minutes to notice that he had a big belly.

"You are observant. Yes, my belly is slightly bigger. It hosts 26 protons and 31 neutrons. As I only have two multiples, it's not easy to find us. But finding the brother with 32 neutrons is even more difficult. You can find only one among four to five hundred siblings."

"It's so difficult to find them! Even so, I will still try my best. See you!"

I got on board the helicopter and left the castle to find those buddies.

"I just introduced the periodic table, now I'd like to talk about isotopes", the guide said, "every castle owner you've visited is an atomic nucleus. Each owner carries protons and neutrons. Those with the same number of protons take up the same place in the periodic table. Among them, those with different numbers of neutrons are brothers. In chemistry, we call them isotopes. So those brothers are isotopes, the most basic members of the periodic table. There are 118 places in the periodic table, each hosts many brothers, or in other words, isotopes. After counting, some concluded that the number of all non-identical brothers is 3473. Take the element iron as an example. Iron takes up the 26th place in the periodic table. In the very place, he already has 29 brothers with different numbers of neutrons."

That's quite a lot of isotopes!

The guide paused for a moment before she said, "now we can draw up a graph according to the number of protons and neutrons of each isotope. The vertical lines represent the number of protons and the horizontal lines the number of neutrons. All isotopes are plotted on this graph according to the number of their respective protons and neutrons."

Now, a blue screen appeared in front of me.

"Look at the graph on the screen. The isotopes gather together from the lower left to the upper right and form a narrow strip with edges slightly bent down. This is the land of isotopes. An isotope is an atomic nucleus or nuclide. And this graph is called the table of nuclides. The blue color in the background looks like the ocean, and the places occupied by the isotopes or nuclides resemble the land. The isotopes or nuclides are sparsely distributed on the upper right side, just like the islands. Having a large number of protons and neutrons, those overweight castle owners are called superheavy nuclides."

"Why are these nuclides distributed only in a narrow strip, rather than in various continents, just like the lands on the earth?" I asked curiously (Fig. 11).

"It's a little bit complex. Let's talk about the protons and neutrons in the castle owners' bellies first. Between each proton and neutron of an atom, there is a nuclear force. Somewhat oddly, this force tends to push the protons and neutrons apart when they are close together, or in other words, when the castle owners press their bellies.

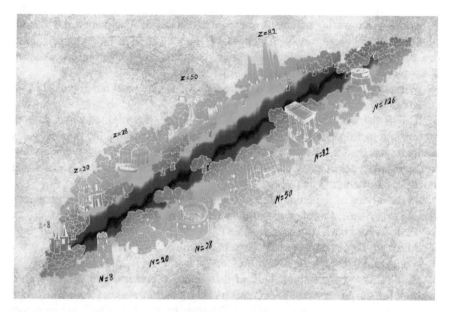

Fig. 11 The map of the dwarfs' world (table of nuclides)

The harder the castle owners press their bellies, the greater the repulsive force will be. Yet it takes tremendous force to pull the protons and neutrons apart, as they stick together like glue, never wanting to part from one another. The greater the distance between them, the weaker the force separating them will be. Besides, the cohesive force between protons and neutrons is far more powerful than that between protons or neutrons."

"Does this force have anything to do with the isotopes?" I asked in perplexity.

"It surely does! The protons and neutrons of the same number will gather together in a more stable manner. So in the lighter stable nuclei, the number of protons and neutrons are roughly equal. Yet because the cohesive force between protons and neutrons is greater, a proton can hold more neutrons. This gives rise to different isotopes. There cannot be too many neutrons, however, otherwise the protons will not be able to hold them. The electrically-charged protons are mutually repulsive, so they cannot stay together with the presence of too few neutrons. This results in a limited number of isotopes."

Accompanied by the guide's sweet voice, I failed some hundred times before managing to find the iron castle owner with 32 neutrons, the fewest among that of his brothers from a multiple birth.

(4)   Encountering Little Friends

We finished our journey in this majestic mountain and flew to another world comprising various mountains and castles. As we flew on, the landscapes were constantly changing, offering a feast for the eyes.

Suddenly, a towering mountain whose peak reached the cloud appeared in front of us, as if blocking our way. Well then, let's embark on an adventure! Our helicopter flew toward it and went inside. With the bright light, I saw a chaotic scene: castles of various sizes sprawled out.

"Where is this place? And which castle shall we visit?" I was hoping to get some guidance.

"This is the world of wood. There are many types of castles, of which carbon, oxygen and hydrogen castles are the main members. Those with the thinnest walls are hydrogen castles. Carbon castles are roughly the same size as oxygen castles, and these two types of castles are slightly bigger than hydrogen castles. So let's visit the carbon castle first!"

On the advice of the guide, we came to a relatively large castle. With previous visiting experience, we found the castle owner before long.

"Hello! May I have your name, please?"

"Make yourself at home! Just call me carbon dwarf."

"Could you talk a bit about your brothers?"

"Sure! I have two siblings, both of them carry 6 protons. The sibling with 6 neutrons has the greatest number of brothers from a multiple birth: 99 out 100 carbon dwarfs are his brothers born at one birth. The other sibling of mine has 7 neutrons. Apart from them, I have a brother with 8 neutrons, coming from another family. He is a troublesome guy. Though being a member of our family, he never settles down even for a while. And before long, he came to other family without us noticing it." The castle owner seemed quite unhappy mentioning this brother.

"How did that happen?" I was confused.

"It's a long story. He was once a member of nitrogen family. This greedy guy stole a neutron and ate it. He then had an upset stomach and vomited a proton. After that, he became a member of our family."

"But how come he became another family's member?"

"He let out an electron! That is to say, one neutron inside him turned proton, so he went back to the nitrogen family again."

"May I visit him?"

"The foster children are so few. You'll have a lot of trouble finding them."

"I will try my luck to find the foster child in your family."

"Well then, wish you good luck."

We said goodbye to this carbon castle owner and set foot on the journey to find the naughty carbon dwarf.

We flew on and kept visiting. Just as the first carbon castle owner had told us, it was no easy task finding the naughty guy.

"One last castle. I'll give up if I still cannot find him." I thought to myself.

We had no trouble finding a carbon castle and its owner. Before I started to talk, I heard a clinking sound and then saw an electron hit the helicopter. I wondered why there were electrons.

"Sorry, don't laugh at me." The castle owner apologized.

"Hello! May I ask what happened just now?"

"I sneezed and let out an electron."

"It's clean here and there are no electrons. How come you inhaled one?"

"I didn't inhale an electron. It's a natural change within me. What a coincidence! Before you came, I was a carbon castle owner. But now I'm already a nitrogen castle owner!"

"How did you manage to change your identity within such a short period of time?"

"We are different from other siblings. We have 6 protons and 8 neutrons in our bellies. With so many neutrons, sometimes we feel slightly uncomfortable. Yet we can stand it. About every 8,000 years, we would sneeze and let out an electron. Spitting out the electron makes us feel extraordinarily comfortable."

"Ah! It's hard enough to find you, though you are no longer the carbon castle owner with 6 protons. But could you tell me something about yourself?" I was so excited that I encountered the change of a nucleus.

"Sure! I used to be a member of the nitrogen family and had 7 protons and 7 neutrons. The universe is filled with all kinds of particles. When these particles travel through the atmosphere, various reactions are provoked. Neutrons, like those in my bellies, are produced during the process. The siblings in our family are all fascinated by the outside world. For example, we may swallow neutrons in the air out of curiosity, and our body temperature would rise as a result. If we cannot stand it, we would spit a proton and become a member of the carbon family. As you have seen, I was still Carbon-14 with 6 protons and 8 neutrons a moment ago." The castle owner paused for a while before he continued, "soon after I became one of the carbon family, I came into two brothers from the oxygen family. We huddled together and floated around in the air. One day, we passed by this castle and decided to settle down here. The average life expectancy of my brothers born at one birth was about 8266 years. That is to say, half of our family members would become nitrogen members in 5730 years of time. But we don't know exactly when this change would happen. It's a real coincidence that you saw me turn into what I am now! I'm part of the nitrogen family again. Without significant changes, I will be Nitrogen-14 with 7 protons and 7 neutrons in my belly for the rest of my life. By the way, there were extremely few members in our family who have 8 neutrons, approximately 1 in 300 brothers."

"What a remarkable story. Thank you for sharing it with me! I hope you will be a nitrogen family member forever!"

I left the castle owner and returned to the helicopter.

"How many ways do nuclei have to transform themselves? Could you tell me more about that?"

"Sure!" The guide said in a pleasant voice. "You are lucky enough to witness the moment when Carbon-14 turned into Nitrogen-14. There exist many ways for nuclei to change themselves.

Such change is called nuclear decay in physics. What you've just witnessed is beta decay ("beta" is usually represented by the Greek letter "β"), in which a negative electron (e^-) is emitted from the nucleus. That's how the change got its name β^- decay. A castle owner would spit a negative electron and a barely visible particle (i.e. a neutrino) if he has too many neutrons in the belly, be it carbon, nitrogen, oxygen, gold, silver, copper, iron or whatever element in the periodic table. During this process, a neutron is converted into a proton. As the castle owner carries one more proton, he would move to the left place next to him in the periodic table. If a castle owner carries too many protons in his belly, he would spit a positive electron and a neutrino. And a proton would be converted into a neutron, causing the castle owner to take one step to the right. This change is called β^+ decay, the principle of which is similar to that of β^- decay. Apart from these changes, a castle owner who has too many protons may take an electron from the walls of his castle and eat it. Then a proton in his belly would thus be converted into a neutron while he spit a small particle (Fig. 12)."

Then I saw a table of nuclides, a graph representing various types of castle owners. There was a curving black line in the middle of the graph, the left side of which were orange and yellow areas and the right was a blue area (Fig. 13).

"Look at the table of nuclides. The black line in the middle represents the stable castle owners who basically remain unchanged throughout their lives. The castle owners on the right, however, would be subject to β^- decay sooner or later; those on the left would be subject to β^+ decay."

"Why is it called beta decay?".

"I'll start with another way of decay: alpha decay. More than a century ago, a renowned scientist named Antoine Henri Becquerel conducted an experiment to further study the X-rays discovered by Roentgen. Becquerel wrapped minerals

Fig. 12 Transformation of the dwarfs: the nuclear decay

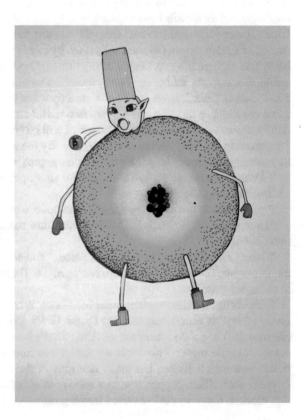

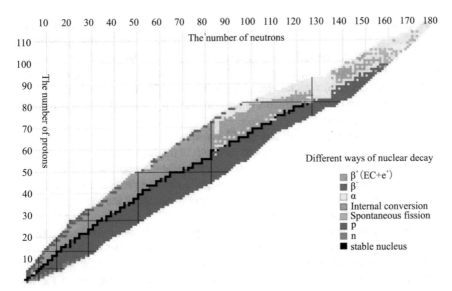

Fig. 13 Table of nuclides

containing uranium and photographic films in black paper, and put these materials under the sun to study if they would emit X-rays after absorbing sunlight. His experiment failed because the sky was overcast during the experiment. After developing the photographic films, he was surprised to find many clear bright spots. He then put these materials between two magnets and found that the rays emitted by part of the materials were deflected in various directions. Those materials emitting deflected rays were electrically charged, and those emitting non-deflected rays were electrically neutral. Becquerel then termed the two reflected rays α-rays and β-rays and the non-deflected rays γ-rays.

Later, a famous scientist named Ernest Rutherford found that the alpha particles (α) are helium nuclei. It means that there is another way of decay: a castle owner spits two protons and two neutrons out. Having lost two protons, he becomes a member of another family. During the process, a helium castle owner is produced. These castle owners are all represented in the yellow area of the graph."

This was a cheerful unexpected discovery.

"Why are there non-deflected γ rays in the middle of the magnets?"

"The helium castle owners are quite stable. They might sway back and forth, but they do so in a rhythmic way. Some castle owners, however, would spin round and round in the slightest agitation or if they ate too much. They would neither slow down nor stop before they release the excess energy. That energy resembling the light wave is γ-ray, yet it is much stronger than the light wave."

"Are there other ways of decay?"

"Yes, there are quite a lot. Take the very heavy element as an example, say uranium nucleus. This guy is potbellied, with 92 protons and 146 neutrons. He has a younger brother with a slightly smaller belly, carrying 143 neutrons. This buddy is very competent as well. He can spit an alpha particle just like his elder brother. Sometimes, he may even divide himself into two fast-running dwarfs and some neutrons. The breakdown without external forces is called spontaneous fission. Yet nuclei are more likely to undergo fission under stress. Take the greedy uranium castle owner with 143 neutrons as an example. Having eaten one neutron, he would split in two and let out two or three neutrons."

"Wait a moment! Is that the fuel used by nuclear power stations?"

"Exactly, the castle owner with 143 neutrons is the fuel used by power stations. But one castle owner alone is not enough. A power station puts several 10^{25} castle owners in a giant stove containing many graphite castle owners and tubes, making them undergo fission within a short time. After colliding with the castle owners, the fast-running dwarfs will let out neutrons.

These neutrons will also collide with the graphite castle owners and end up being eaten by some greedy guys. After that, the greedy guys will have a high fever, with their body temperatures rising to hundreds of degrees Celsius. The water in the tubes is used for cooling down these graphite castle owners. Because of the high temperatures, the water will turn into steam which then produces electricity."

"So this is how a nuclear power station works!"

"But I digress. Let's talk about other ways of decay. Having spit a positive electron, some castle owners on the far left of graph may still have an upset stomach. In this case, they would let out a proton (β^+delayed proton emission). Some may just spit a proton (proton emission). A small number of castle owners who feel terribly uncomfortable for having too many protons may let out two protons. These three ways of decay only occur in the nuclei on the far left. The castle owners on the far right have many neutrons which make them feel quite uncomfortable. Like those on the left side, some of these castle owners may spit a negative electron, and then let out a neutron (β^-delayed proton emission); some may just let out a neutron (neutron emission)."

These castle owners have quite a few tricks!

"But more than that! Like I said, some heavy nuclides would divide themselves into two castle owners; some may let out a negative electron and then divide themselves in two, the process of which is called β^-delayed proton emission. Some heavy nuclides castle owners may not only give birth to a little α castle owner, but also a slightly bigger one, say Carbon-14 (with 6 protons and 8 neutrons), Oxygen-20 (with 8 protons and 12 neutrons), Neon-24 (with 10 protons and 14 neutrons) or Magnesium-30 (with 12 protons and 18 neutrons). Because these parent castle owners have many neutrons in their bellies, their newborn children also carry many neutrons. After giving birth, the castle owners will slim down and have fewer protons in their bellies. And they will be rearranged in the periodic table accordingly. But these are rare phenomena, which can only be seen through special devices in the laboratories dedicated to nuclei research."

These castle owners are amazing! I do need to study hard and broaden my horizon. But now, let's call it a day. See you next time!

2 Collision Contests Between the Dwarfs and the Castle Owners

Could Dwarfs Hit the Targeted Castle Owners?

The dwarf soldiers from the training site flew at a rapid speed. To see more clearly the collision contests between the dwarfs and the castle owners, we used the ultra-high-speed camera in the helicopter to film the whole process of collision and then replayed it with a super-slow-motion camera.

The dwarf army dashed forward as if they had entered a place where there was no one. Moving so fast, they didn't have time to look around. They just kept running forward. It was so difficult for the dwarfs to hit the castle owners with one attack (Fig. 14).

Let's do some calculations to see how likely a dwarf flying straight is to collide with a target castle owner. Let's say the target is a dartboard having a diameter of 20 m and the bull's eye is only two square millimeters.

The 10th ring is a circle with a radius of 1 m and the bull's eye being the circle's center; the radius of the 9th ring is 2 m, and so on. The radius of the 1st ring is 10 m. Without a sighting telescope, a dart from 1 km away is equally likely to hit any point of the target. The probability of the dart hitting the 1st ring is the ratio of the area of the 1st ring and the area of the whole dartboard, equaling 19%. Calculated in

Fig. 14 The dwarfs hit the nucleus castles

this way, the probability of the dart hitting the 10th ring is 1%. There is a 1 in 100 million chance that the dart can hit the dartboard's center, namely, among 100 million darts, only 1 dart can hit the center of two square millimeters. So there is a 1 in 100 million chance that a dwarf can hit a target castle. But a dwarf travels long distances through the targets. For example, if a dwarf passes through 10,000 randomly-located castles, the probability of him hitting the castle owners will increase by 10,000 times. Besides, there are numerous soldiers in the dwarf army. When the army pass through the castles, there is bound to be many collisions. It will become more interesting if the soldiers' hands touch that of the castle owners, or if the soldiers' bodies touch the castle owners', or if the soldiers straightly bump into the castle owners. Various things can occur in different situations. Now let's take a look!

(1)   The First Encounter

Every dwarf in the army carried a strong positive electron; each castle owner had a positive electron to protect him as well. When these little soldiers approached the castle owners at the speed of lighting, the castle owners would build a line of strong defense with their positive electrons to prevent the dwarfs from coming close. But if the dwarfs came too fast, the defense was of no help at all. At the distance, a dwarf attempting to approach the castle owner felt an invisible repulsive force. So he had to bypass the castle owner. Another dwarf who ran into the defense was pushed away by a powerful force. After observing for a while, I noticed that the dwarfs would be pushed farther away the closer they came to the castle owners. But then, a strange thing happened. A little soldier bumped into the castle owner. However, he didn't bounce away, but rather ran around the castle owner for a moment before he ran away (Fig. 15).

I thought about it and came to understand the reasons behind. The little soldiers wore armors with positive electrons, and the castle owners wore the same armors too. Like charges repel. The closer they get, the greater the repulsive force will be. If the little soldiers were 10 m away from the castle owners, there would be 1 unit of repulsive force. Should their distance from the castle owners be 5 m, there would be 4 units of repulsive force. If they were 1 m away from each other, the repulsive force would rise to 100. There existed, however, another force between the little soldiers and the castle owners which resembled the attractive force between opposite poles of two magnets. Such force came from the small balls in the dwarfs' bellies. Strangely enough, this force only worked within a close range. As the dwarfs came closer, this force would increase very fast. But there would be no such force if the dwarfs were farther away. So that explained why the castle owner pushed the little soldiers away fiercely when they were far away, and gently pushed them when they were close. That's why the little soldier could ran around him without being pushed away hard (FIg. 16).

There was another situation. If a little soldier held hands with a castle owner, the little soldier would slow down and the castle owner would move one step forward. In this case, the castle owner would climb up either one or several steps, or the little soldier was pushed up one or a few steps, or these two guys climbed up one or two

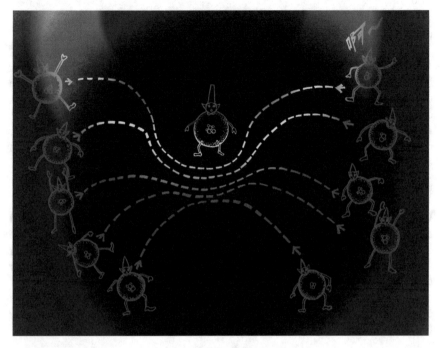

Fig. 15 The faraway hills are green (the magic nuclear force)

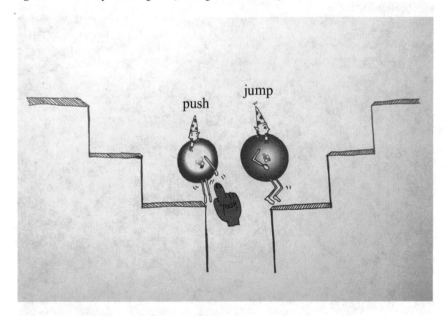

Fig. 16 The dwarfs jumping the steps (radiation)

steps. Before long, the two guys would jump down and emit some invisible rays, i.e. the γ-rays.

When they first met, the little soldier and castle owner were both too shy to talk with each other. They just started to exchange presents. The presents were neither flowers nor rings, but rather something more lavish—the neutrons and protons in their bellies. After the exchange, something unexpected happened. These two guys changed their appearances completely. The once little soldier became a passerby, and the once castle owner took a new look. Both of them may jump to a higher step, but before long, they would go downstairs and live their own cozy lives because of the emission of γ-rays.

(2)   The First Fight

Like a super rocket, an argon little soldier with 18 protons and 18 neutrons headed for the 8th ring of a target beryllium castle. With only 4 protons and 5 neutrons, the beryllium castle owner was way smaller than the little soldier. So the castle owner started to retreat at the sight of the little solider in the distance. As the castle owner didn't move as fast as the dwarf, they crashed into each other with a bang.

When a bullet hit a rock, both the bullet and the rock would be damaged. It was also the case when the argon little soldier collided with the beryllium castle owner (Fig. 17).

Fig. 17   The battle between atomic castle owners (transmutation)

Knowledge Link

Speed

To send an artificial satellite into space and make it orbit the earth, the speed of its carrier rocket is required to reach 7.9 km/s. If the satellite is to leave the earth and orbit the sun, the required speed is 11.2 km/s. If it is to leave the Milky Way, the required speed is 120 km/s. The speed of light is 300,000 km/s. The speeds of the little soldiers are between 35,000 and 290,000 km/s, which are way faster than that of an average rocket.

The two ran into each other at the 8th ring. It was likely that they would both get hurt. In an eye blink, the little soldier argon looked refreshed, turned into a new solider and kept moving forward, followed by some scattering neutrons, protons and other things. These things might had been spit out by the argon or the castle owner because of the squeeze in the hit (Fig. 18).

Getting hit, the castle owner felt dizzy. He rapidly retreated the spinning before he managed to stop. When I took a closer look, I found that he had already transformed himself into another castle owner.

Who was this new soldier? To confirm his identity, we needed to find out the number of protons and neutrons in his belly first. So fast did he run that we needed a more accurate timer whose resolution was 10^{-11} s. It took the previous little soldier 57.1 ns (1 ns equals 1 billionth of a second) to run 2 m at 35,000 km/s. Apart from measurement of the time, the speed of the new soldier passing through some matter at certain width was also required to find out how much energy he lost in the process. The quantity of the lost energy and the soldier's speed were both related to the ratio of the number of protons to the total number of protons and neutrons. To perform this measurement, it only required a detector made of silicon chips which were not too thick.

If you arrange the new soldiers according to the quantity of their lost energy and the time they spent on covering identical mileage, you'll find them perfectly aligned. Those standing together were brothers born at one birth, who kept a distance with other siblings. Those who could be connected by an imaginary line were from the same family. The brothers from a multiple birth were mostly the previous soldiers. These soldiers did take part in the collision game, but they missed the chance to meet the target castle owners. So these passersby remained unchanged throughout the process. On their left stood the brothers who had their neutrons ejected during the collisions. The brothers on the right were luckier: they grabbed a neutron in the collisions. Below the argon family was the chlorine family, followed by the phosphor family... You could tell from the alignment that many brothers from other families were born during the collisions of the argon dwarfs and the beryllium castle owners. Many members from various families had been produced with the approach of collision. Though short-lived, these members' longevity and other features were

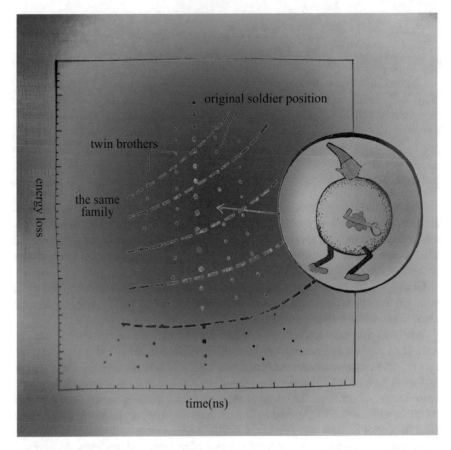

Fig. 18 Taking seats by the order of their families

discovered by scientists in various ways. Indeed, the collision approach has made its fair contribution to producing more nuclides (Fig. 19).

(3) A Loving Hug

A dwarf soldier who didn't run fast enough accidentally bumped into another dwarf. The two guys then huddled together, jumped up onto a slope and started to dance. So happily did they talk to each other that they no longer wanted to be apart. After a short secret discussion, they decided to merge themselves into a stout dwarf. This merged dwarf kept spinning while his body temperature rose significantly. So exhausted was he that he nearly fainted before managing to ease the burden of his belly. He let out some neutrons and two protons, and then gave birth to a baby helium. While he was vomiting, he emitted γ-rays and skipped down the slope. At every step, he emitted the γ-rays which were not visible to the naked eye. In the end, this stout dwarf turned into a handsome guy. However, he did not remember anything of the change.

Fig. 19 A loving hug (an inelastic collision)

What happened in other loving hugs? A couple were somewhat far away from each other. They just held hands before they were separated unwillingly. Another couple kissed and exchanged protons and neutrons as keepsakes. Their appearances were changed subsequently. This new couple then part with each other reluctantly.

(4)   The Fierce Battle

A dwarf soldier rushed forward at the speed of light. Whoever saw this would feel a bit scared. The little solider hit a castle owner in his first attack. Though much fatter than the little solider, the castle owner could not stand the impact of the hit (Fig. 20).

In the blink of an eye, the little soldier got into the castle owner's belly. Just like the Monkey King, he turned the castle owner's belly upside down. No matter how hard the castle owner begged for mercy, he just kept stirring the belly. The castle owner had to let out neutrons, protons and other little castle owners, or anything to make himself feel better. The more he vomited, the smaller his belly will be. After a long time, the castle owner finally settled down. He looked around, finding lots of scattering neutrons, protons and castle owners of various sizes. The castle owner became much thinner, but he totally forgot what he had looked like before the hit.

(5)   The battle between a greedy fat guy and the neutrons

A neutron was wandering in a uranium castle. The walls of the castle could not block the neutron and electrons were bounced away at the touch of it. But when it accidentally bumped into the castle owner, it ricocheted and vanished into the air right away, just like what would happen when a marble hit a giant rock. The uranium

Fig. 20 The fierce fight between the dwarfs and a castle owner of nucleus

castle owner was incredibly greedy. With a belly about to burst, he still wanted to eat something, as though he hadn't eaten for days. He was particularly hungry for neutrons which was his favorite. So whenever a neutron came across him, he would open his mouth to eat it. Because of the fat belly, however, the castle owner became sluggish. Most neutrons had already run off when he opened his mouth. So he usually could not enjoy his favorite food unless a neutron hit right on his mouth.

He had been long waiting before a neutron suddenly hit his chin. He lowered his head, opened his mouth and swallowed the neutron in no time. He then licked his lips and smiled in contentment. Before long, his broad smile turned into a bitter one. Turns out his belly bulged into the shape of an enormous pumpkin. The "pumpkin" was then reduced to a "dumbbell", the middle of which was narrowing down. After a while, this "dumbbell" became two little baby dwarfs connected by an "umbilical cord". Boom! The "umbilical cord" broke down and the two dwarfs ran fast in opposite directions, seemingly for fear of being bound together again. There were some neutrons rushing out as well. The two dwarfs had enormous energy. The combined energy of them reached 200 meV (mega-electron volt). If they were stopped, the energy would be transformed into heat (Fig. 21).

(200 meV = 7.66×10^{-12} Cal (calorie); A calorie of the temperature of 1 g of water by 1 °C.)

This is called fission which occurs after nuclei absorb the slow neutrons (thermal neutrons). Nuclear power stations generate electricity with the heat produced in the

Fig. 21 The transformation of the super-fat guy: nuclear fission

continuing fission. An explosion will occur if enough castle owners that undergo nuclear fission after swallowing thermal neutrons are gathered together, coupled with some guys undergoing fission without eating thermal neutrons. This is how atomic bombs are made. Oh, so changeable and complex are the dwarfs' collisions!

3 A Special Performance

I had no idea how long it took before we finished our trip. When we got out of the helicopter, we returned to our original sizes. I came to see a special auditorium of a research institute. There was no large screens or exciting music, not to mention crowded audience. One could only find rows of equipment connected by many neatly-arranged black and red wires. Some young teachers were working with an elder tutor, and the graphs on the computer screen were in constantly change. The tutor held an exquisite screw driver, adjusting the equipment carefully and then looking at the graphs on the screen. The process repeated for a while before he stopped with a satisfactory result. I took the opportunity to come up and ask: "Excuse me, sir. What are you studying on?"

"The forms of the atomic castle owners", said the tutor briefly yet patiently.

"Aren't atomic castle owners like tiny balls? How come they take different forms?" I was quite surprised.

"Not really. Only part of them are round, most are in various forms. Now I'll show you their forms roughly on the computer."

"Wow, that's wonderful!"

Fig. 22 Main forms of the dwarfs

The tutor had me sit in front of a computer. He exited the page with graphs and hit some keys. On the screen appeared a little castle owner with a round belly. The tutor pointed at the castle owner, saying: "Only the castle owners with protons and neutrons in magic numbers have round bellies. The confirmed magic numbers are 2, 8, 20, 28, 50, 82 and 126. Take Lead-208 as an example. He has 82 protons and 126 neutrons in his belly. Both 82 and 126 are magic numbers, so he has a round belly resembling watermelon."

As he was saying, he hit the keys again. There appeared another castle owner with a longer belly, which looked like a cantaloupe. The castle owner kept turning somersaults. He was just like a performer of the acrobatic fighting in the Peking Opera!

"Look, this castle owner's belly is longer. You see, he keeps spinning. Actually, that castle owners with round belly were also spinning."

There appeared a pumpkin on the screen. The tutor said with some hand gestures: "Some castle owners expand horizontally and their bellies look like pumpkins. These castle owners can do somersaults as well. Sometimes they can even move quickly!"

"Ah! Is this a banana?" I couldn't help but shouted.

"Yes. When some castle owners move very quickly, their bellies will become as long as bananas! This form was discovered more than two decades ago."

After a flash screen, another castle owner appeared in front of us. He looked odd. His belly was constantly swelling and contracting, just like what happened when one

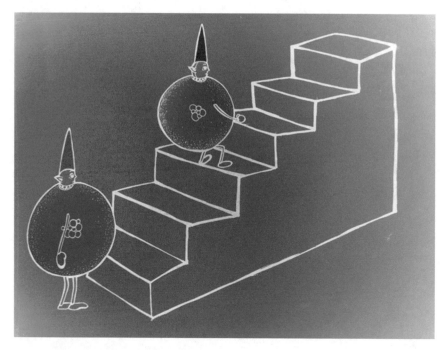

Fig. 23 The energy staircase of the dwarfs

took deep breaths. The belly was first bigger on the top and smaller at the bottom. In no time, it became the other way around. That's awesome! (Figs. 22 and 23).

"These castle owners don't have constant forms. They have been taking deep breaths since they were born."

"Are there any other forms?"

"Quite a lot. Look!" The tutor said, hitting the keys again.

There appeared something like kiwi. It was in constantly change, taking the form of pumpkin first and then kiwi again. For a while, it didn't change its forms, but rather spun around at a great speed. Oh! It changed its form again! It was spinning as well as swaying.

"How can you identify their forms and the ways they change?" I asked in perplexity.

"It's a long story. In short, various forms and movements represent different energy bands, like various energy staircases. If the castle owners are suddenly placed on top of different staircases, they will go down the stairs step by step. With every step they take, they will emit a γ-ray. We can tell their forms and state of movements by measuring the changes of these γ-rays' energy. To put the castle owners on various staircases, we adopt different approaches. The main approach is making various dwarf soldiers collide with the castle owners. Now we are measuring the energy of these γ-rays."

"So amazing! I'll try to learn more about it. Thank you for your explanation!"

After the visit, these animations were still lingering on my mind.

Training Bases of the Dwarf Troops

Guoqing Xiao and Xiaodong Yang

To train competent combat dwarf troops, scientists set up various training bases, big or small. How do the training bases look like from the inside?

1 Why Are the Dwarf Troops Set up?

To fight. Fighting against the enemy is a fight, rescuing and providing disaster relief are a fight, and even endeavoring to innovate is a fight. In the dwarfs' world, there exist many puzzles to solve and difficulties to overcome. For example, how vast is the territory of the dwarfs' world? Where are the dwarfs from? What's the weight of the overweight guys? How many "enemies" are to be defeated? To be more specific, when the cancer cells invade people's bodies, how can the dwarf troops provide assistance? There are a lot of innovative causes as well. The dwarfs have to think about how to make flowers more beautiful, the crops more productive, and so on. All these require the dwarf troops to fight and complete.

2 How to Build Dwarf Troops?

It is not easy to build a combat force. First, qualified fighters should be selected to undergo long-term rigorous training, which is carried out at a training base. Then, what are the training bases for the dwarf troops like? It is said that the training bases of the dwarf troops can be divided into many kinds, and different training bases are

G. Xiao (✉) · X. Yang
Institute of Modern Physics, Chinese Academy of Sciences, Lanzhou, Gansu, China
e-mail: xiaogq@impcas.ac.cn

© Zhejiang Education Publishing House 2021
G. Jin and G. Xiao (eds.), *Probing into the World of Nuclei*, China's Big
Science Facilities, https://doi.org/10.1007/978-981-16-0715-8_2

Fig. 1 High voltage multiplier

constructed according to various combat tasks so as to train different combat units (Fig. 1).

Each training base has the main departments such as the Recruitment Department, the Troop Corridor, the Speed Training Department, the Combat Target Department and so on. Most training bases have only one Speed Training Department, and a few large ones have two connected training departments. Speed training is to make the troops faster and more capable of fighting. Venues of these training bases should be rid of as many as other castle owners, with a maximum of only 10^{-13} left. This is called an ultra-high vacuum environment.

3 Humble Dwarf Troop Training Bases

Well. Let's look around. First of all, let's board a helicopter made of special materials and fly straight to the Multiplier training base for dwarfs (Fig. 2).

The helicopter hovered before a huge building where many tall pillars were connected. It was the earliest dwarf training base. The big pillars were insulated and the voltage level above was higher than the first class. Different methods could be used to increase the voltage, including doubling, transformer and discharging. The Recruitment Department of the dwarfs was located on the top floor. Its task was to activate the castle owners, to make the castle owners move quickly, and to knock off some of the electrons on the walls, thus reducing the castle owners' burdens and making them carry positive charges. There were many ways to get rid of the electrons. The simple one here was to take two metal bars, connect them to the positive and negative poles of the great power supply, and put them together with the dwarfs'

Fig. 2 Arc light

castles in the middle, and then the electric arc would appear between them. In doing so, the electrons on the walls of the dwarfs' castles were knocked off.

There were magnetic fields around certain discharging areas to imprison the electrons that had been knocked off and prompt them to knock off other electrons on the main castle walls. By these means, a lot of soldiers with positive charges could be recruited. These soldiers were located at very high potentials, as if they were on the high mountains. At an increasingly faster speed, they ran through the passage of the troops towards the hills till they reached the targets.

Such training bases were mostly used to train the slim dwarfs in the early days, such as the hydrogen and deuterium dwarf forces. There was only one electron on the walls of these castle owners. After the electron being knocked off, the owners were all relaxed and carried a positive charge, and they ran from the top to the bottom of the mountain at the same speed. The dwarf soldiers from these training bases couldn't run very fast. Most of them ran at about 1/20 of the speed of light, i.e. a little more than 10,000 km/s (Fig. 3).

The helicopter flew to a small training base. It drilled into a large round tube filled with high pressure insulating gas. There stood a hollow long column which had a large round metal head, evenly inlaid with bright metal rings that were connected to a string of red sticks (resistors) outside. Attached to the column was a wide conveyor belt. That belt was an electricity transmission elevator which continuously passed electrons from underneath to the top to make the voltage on the top a few million volts. The belt then evenly divided the voltage to the metal rings on the long column through a string of red sticks below, and the long column became a training field for the dwarf troops. The Recruitment Department of Soldiers was right on the long column. Of course, the soldiers were all positively charged (Fig. 4).

Wouldn't it be too wasteful to use this high voltage just for one time? Was it possible to use it once more? Good idea! So on the top of the big head, another piece of hollow long column was connected, with its top attached to the ground. A

Fig. 3 Training ground of the dwarf troops

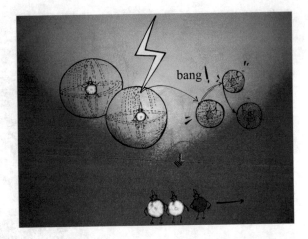

Fig. 4 Schematic diagram of high voltage multiplier

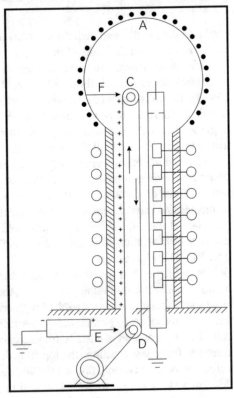

thin metal film was added between the two columns. In this way, one could form a training ground in the shape of "hill" (ground – positive high voltage – ground). The Recruitment Department of Soldiers was placed on the "ground" outside the steel cube, and then the soldiers were sent out through the passage. How was training done here? First, the selected soldiers were required to carry one more electron. When they came here from the outside, they would be absorbed to the "top of the hill", then ran quickly through the film in the middle of two long columns, unloaded some of the electrons on the castle, carried some positive electrons before they rushed down the hill to further improve their speed. The training base connected by the two columns had a special name called "tandem accelerator". The troops from here could reach the speed of around 30,000 km/s, way faster than that of rockets. Such an accelerator may stand on the ground, but more often it was placed horizontally because of the height.

Have you also thought of other ways to make full use of the precious high pressure to train the dwarf troops more effectively?

4   Training Base of the Fat Dwarfs

Coming out of the steel tube, we flew straight to a hall. Its walls were made of cement over one meter thick. In the middle stood a tall square iron frame with a thick circular pole on each end. A stainless-steel box of the same size was sandwiched between the two poles. This was the indispensable equipment of the dwarfs' training field—the magnet system. There lay a very thick stainless-steel tube dragging a long tail with many water pipes. This tube was called a resonant cavity. It matched the two yellow pipes from above the big iron frame to send high-frequency electric power to the training ground. Two large vacuum machines stood on both sides like two bodyguards, holding a stainless-steel box tightly in fear of being robbed of. The task was to clean up the training field (Fig. 5).

There were two yellow pipes above the big iron frame responsible for sending high voltage at high frequency into the training ground. This training base was complex. But the real training ground was inside the stainless-steel box, the outside of which were all the security facilities. This was the modified cyclotron, code-named SFC.

Let's go in and have a look. Hold on! You can't have a bit of iron on you, otherwise you will be absorbed in the magnetic head. It was empty. Except for the poles like three fan blades and some copper tubes coiled on them, there was only a semicircle purple copper boxes connected to the guy lying on the outside, and a piece of purple copper attached to a yellow pipe on the outside. Right opposite to the straight flange of the semicircle copper box was a rectangular copper frame, connected by another yellow pipe. In the center of the copper box was a small tube that led to the Recruitment Department below. Near the edge of the empty part was an arc channel separated with stainless steel.

The Recruitment Department was larger, more complex, and more specially designed than the previous one. There were super conductive coils (superconducting

Fig. 5 Lanzhou heavy ion research facility—sector focusing cyclotron (SFC)

coils) arranged in the shape of the special magnetic field, special "wireless charging" source and the extreme cooling devices of minus 270 degrees Celsius. The department recruited a wide range of soldiers, from the slim hydrogen, lithium dwarfs (with 3 protons and 4 neutrons) to the fat uranium. The soldiers here were more capable. With less electrons, most soldiers carried a light burden. Some did not even have any electrons. How was that achieved? To put it simply, the lazy (solid) castle owners were "placed" in a specifically-designed small pot which was heated to move them away from each other (gas state). And the active castle owners would be sent in directly. At the same time, a rapid electronic force was set up to continuously provide the troops with energy through the wireless charging device, so that the soldiers would always have remarkable fighting capacities to attack the walls of the castles and knock off the electrons there. With the special design of the intense magnetic field, the castle owners, whose walls had been pulled down, would be put in a small space, guided to the exit in the electric field and sent to the training fields through the passage. Sometimes it was necessary to gather these soldiers again somewhere in the passage and send them to the training ground in batches. Such recruitment department is called ECR ion source.

In order not to obstruct the normal training of the dwarf troops, the gas castle owners in the stainless-steel box were almost driven out by the vacuum machine. After coming up from the transport passage in the center of the copper box, the positively-charged dwarf soldiers began to run driven by a powerful magnetic force and was charged every time they passed by the straight flange of the semicircular copper box. Gaining more energy each time, they speeded up a bit more. In this way,

Fig. 6 Schematic diagram
of the cyclotron

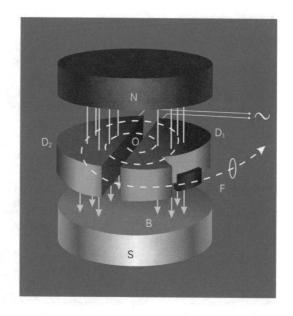

the dwarf soldiers ran faster and faster, but the time used for each lap was equal. Having crossed the outermost arc passage, they entered a new troops passage. The troops trained here were even faster, and the speed of the carbon troops was about over 50,000 km/s. More importantly, all dwarfs could be trained here regardless of their weight. Of course, the heavier they were, the slower they ran. For instance, the speed of the Uranium^{-238} dwarf was less than half that of the carbon dwarfs. This training base was called cyclotron (Fig. 6).

Could dwarfs be trained to run faster? Absolutely. But this training base was not capable enough, for its equipment was not advanced enough, and it was not big enough neither. Let's go to a larger training base!

The helicopter made several turns in a narrow corridor and came to a hall. Ah! What a big guy. Each of the four huge iron "heads" occupied one corner of the walls. Each of these four magnets weighed 500 tons. They were collectively holding a large flat stainless-steel tube, and inside the tube was the training field. On one side, two transport passages were connected to the tube; one extending to the distance and sending here the troops that needed further training; the other reached the next hall, from which the trained troops would go to the battlefield. On the other hand, the high frequency cabinets, which was to continuously provide power for the trained dwarfs, stood on opposite sides.

Sitting in the middle of the four magnets was a large vacuum machine, which was responsible for clearing away outsiders in the training field. This was also a sector separated cyclotron (SSC) (Fig. 7).

Let's go inside and take a look. Turn smaller!

The helicopter turned into the size of a dwarf castle and passed through a small hole into the flat stainless-steel cylinder. What a wide space! There were several

Fig. 7 Lanzhou heavy ion research facility: sector separated cyclotron (SSC)

short passages near the center, and two enormous mouths opening in the opposite as if ready to bite something at any time. There was an isolated passage near the edge. A lot of dwarfs were seen whizzing in from the outside, one group after another. Each group of the dwarf troops would go into the training ground through the short passage near the center, and began to run in turns, in arc lines and then in straight lines, along the line-free runways. They were subjected to electric shocks every time they passed by one big mouth. As a response, they jumped out a little bit and changed to another track and ran faster. The entire training ground was full of troops running at different speeds in an orderly manner.

Only a small number of laggards wandered aimlessly and finally bumped into nowhere. After some hundreds of laps, they finally came to the edge of the wall. Most soldiers of the troops went along the outside of the wall, yet a few poor guys didn't see the wall before they hit it. I bet you know what would happen following the hit. In fact, it was possible to train a variety of dwarf troops, be it a small carbon dwarf or a fat uranium buddy. They could all be trained into qualified soldiers. The troops sent out from here ran faster. Yet inevitably, some soldiers would be eliminated, so the number of the soldiers would be smaller than that when they first came here (Fig. 8).

We continued to walk along the troop passage to review the dwarf troops. In addition to a few open gates, there were several small copper tubes hung up along the long passage. The dwarf troops moved forward in groups, keeping the same distance between each two groups. Each group of the dwarf soldiers was not as neat

Fig. 8 Training scene of the dwarfs: the working principle of SSC

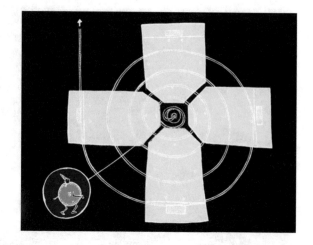

and uniform as the guard of honor. It was more like a procession, moving forward freely, sometimes parting and sometimes gathering.

The rare soldiers who were too far away or who didn't keep up with the pace of the group could no longer return to the group, because they had no idea where the wall of the transport passage was. In fact, the gathering of the troops was controlled by the magnetic force of the quadrupole magnets outside the passage, and dipole magnets were placed at the curves and outside the passage to force the troops to take the curve.

To meet the needs of special combat, some dwarfs (new particles) were also produced, or a new substance, the quark gluon plasma, was produced at super high temperatures. The temperature of this substance reached over 1,400 billion degrees. It was said that this substance existed only in the initial stage of the big bang. This required increasing the speed of the dwarf troops to nearly the speed of light — 299,792.458 km/s. Of course, the dwarf troops were unable to reach the speed of light. Because when the dwarf moved very fast, its mass would increase rapidly. The closer the dwarf moved at the speed of light, the greater its mass will be. For example, when the speed was 99% times that of light, the mass was 7 times as much as the original; when it reached 99.9%, the mass was 22 times as much as the original. This was also the reason why the speed of any object could not reach that of light. Therefore, to make the dwarfs run faster, a larger new training base—synchrotron was needed.

5 Dwarf Lightning Unit Training Base

Let's have a look at the training field of synchrotron (cooling storage ring CSR). The helicopter came in from the door and turned a few corners, and suddenly there was a stainless-steel pipe which extended to both ends and was surrounded by many iron

Fig. 9 Lanzhou heavy ion accelerator cooling storage ring (HIRFL-CSR)

blocks of different colors. The training base was a special circular site surrounded by a thick stainless-steel pipe. The circumference of the inner side of the pipe was nearly 162 m. Before the eyes, a longer hexagonal metal tube tightly wrapped the pipe. Taking a closer look, I found the tab read "high frequency acceleration cavity". It turns out the tube was the special equipment used to stimulate the dwarf soldiers and to increase their speeds. As I walked to the left along the pipeline, the first thing I saw was a piece of yellow iron that read "quadrupole magnet". (Figs. 9 and 10).

This was the main equipment to restrain the troops. Together with other magnets, it constantly revised the direction of each soldier, and kept the whole troops consistent in the course of training. There were 30 "quadrupole magnets" on the base. Beside the magnet laid a guy connected with the pipe and wrapped in white metal cloth. After consulting the management staff, we learned that this was the "vacuum pump", which was used to clear the idle dwarfs in the training ground. The site required less idle personnel, probably $1/10^{15}$ of the average air. Hard work must be done to meet this requirement: not only installing a number of advanced "vacuum pumps", but also making a tube with special stainless-steel material. The tube was specially treated. Before the work, it was baked and cleaned at a temperature of more than 200 degrees for dozens of hours. According to statistics, there were dozens of "vacuum pumps" on the base. We went on and saw a long piece of big blue iron, four blocks in a row at the corner. There were evident tabs reading "dipole magnets". We knew that the dipole magnet was to force the dwarf troops to turn, and there were 16 dipole magnets responsible for this here. Further on, some quadrupole magnets and dipole magnets were alternately arranged. Suddenly, two sturdy pillars were lined with a

Fig. 10 Lanzhou heavy ion accelerator cooling storage ring: electronic cooling device

long stainless-steel pipe above the stainless-steel pipe, and the thick pipe was covered with orange yellow square equipment. This was a new thing. We decided to consult the experts.

The experts told us that it was the "electronic cooling device" used to cool down the dwarf troops. What was electronic cooling? Let's first talk about the temperature. Temperature is the velocity of free movement within a body of objects (atoms or molecules). If the river flows, the water molecules move in one direction. Likewise, this movement has no effect on the temperature. Only a free movement can make the temperature change. The water molecules in boiling water move very quickly, and the temperature is very high. If some cold water is added to the boiling water, the temperature will drop immediately. This is the cooling phenomenon. By colliding with hot water molecules, the cold-water molecules slow down the rapidly moving molecules, and they themselves move quickly, and at last the speed of the free movement of all is almost the same. The dwarf troops seemed to be moving forward in one direction. But if you went into the troops, you would find that the troops were not as orderly as the guards of honor. In fact, every dwarf soldier in the group was free: they could walk a little faster, or a bit slower; they could walk to the left or the right side, so that the dwarf troops would maintain a certain temperature.

But this would influence the fighting capacities of the whole troops! They must be rectified, but they could not stop. What was the solution? A powerful group of nearly ten million trillion electronic soldiers per second, the speed of which was as fast as the dwarf troops, was put in to carry the dwarf troops comprising less than 1/1,000 times the number of the electronic soldiers. In this way, if the dwarf soldiers drifted

Fig. 11 Reducing the
temperature of the dwarfs:
the working principle of the
electronic cooling device

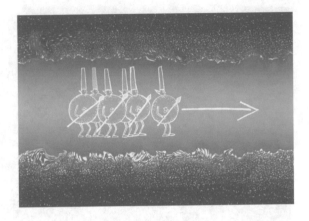

this way and that, they would encounter electronic soldiers. Although the electronic soldiers were small, their number was large. The more of the collisions, the more obedient the dwarf soldiers were; they could be as fast as the guards of honor. After a period of supervision, the electronic troops left the dwarf troops and then returned to their camps. After that, new electronic forces continued to join so as to ensure that the dwarf troops would be rectified and become a qualified combat force (Fig. 11).

The training here was divided into three stages. The first stage was gathering the troops. That was, to gather and group the soldiers who had traveled long distances from SSC or other former training bases. After a certain number of soldiers were gathered, say, one hundred billion carbon dwarfs, a new training would start. The second stage was training that lasted for a few seconds in general. The soldiers kept running on the only one track, scattered and gathered repeatedly. However, under the restraint of the "quadrupole magnets" and other magnets, they would move along that track as much as possible. Each time they passed the "high frequency acceleration chamber", the dwarf soldiers would be stimulated, and their speed was thus improved. At the same time, the magnetic field of the "dipole magnet" should be increased correspondingly, so that the soldiers would not be thrown out at the bend. When their speed was up to standard, the troops would be sent to various battle fields to fight. The third stage was transporting troops, a tricky process as well. There were two ways. One was to send the qualified soldiers out of the track group by group. This was called slow delivery. At the slowest pace, it took ten thousand seconds to send out all the soldiers. Another way was to send all the soldiers out of the track at once, which was called fast delivery.

The soldiers trained here had a greater speed. For example, the speed of the carbon soldiers could reach about 90% that of light, namely 270,000 km/s. The fight of these dwarf soldiers could not only help new dwarfs to be added to the table of nuclides, but also kill cancer cells, fight with the DNA molecules of plants and transform DNA to produce new varieties of crops, and make the flowers more colorful.

There were bigger training bases, such as the RI HC. Its track was 3,850 m long and the soldiers trained were even faster. In Europe, there was a Large Hadron Collider

Fig. 12 The European Nuclear Center: The Large Hadron Collider (LHC)

called LHC, whose track was 27,000 m long. The speed of the dwarf soldiers it trained was 99.9998% of the speed of light, which was 299,791.9 km/s. The dwarf troops trained by LHC collided with each other and produced the long-awaited "God" particle, the Higgs boson. It was much heavier! Its mass was about 126 billion times that of electrons, or 68.6 million times that of protons. This type of training base could not only train the dwarf troops, but also train smaller and faster electronic troops. Electronic troops could also perform their abilities in many battlefields. If the negatively-charged squadrons collided with the positively-charged squadrons at near the speed of light, all kinds of new particles would emerge.

The extremely high-speed electronic troops would send out X-rays at the curve, which could be used to examine the bones in small fish and to see the internal structure of various proteins. If you have the chance, you can visit such training ground in Beijing and Shanghai (Fig. 12).

OK. Enough for introduction this time. Goodbye!

New Members of the Dwarf World

Zaiguo Gan, Zhiyuan Zhang, and Jinjie Liang

How many inhabitants are there in the dwarf world? Where do they come from? How do they come to the dwarf world?

1 Map of the Dwarf World

The nuclide chart is a map of the dwarf country, or the map of dwarfs in the ocean made up of numerous protons and neutrons. In this chart, the longitude is the number of neutrons and the latitude is the number of protons. In the world of dwarfs, every dwarf has a definite address and remains independent inhabitant. On the basis of the number of protons and neutrons in the dwarf's belly, you can accurately find address of the dwarf you are looking for. As the dwarf at each location represents a nuclide, this map is named a map of nuclides. Currently, about 3,473 dwarfs are found inhabiting in the world of dwarfs.

The various dwarfs on the twisting black line of the chart are very stable. Unfortunately, "people" (nuclides) as stable as such are not so many, only about 288 in total. Among them, 254 live in longevity; 34 are particularly long-lived. Their life expectancy generally exceeds 80 million years; some exist as long as the earth (4.5 billion years) or even longer, such as ^{232}Th, ^{235}U, ^{238}U etc. The rest 3,000 inhabitants are unstable; some are long-lived to about 10,000 years, yet some short-lived to just several nanoseconds (several out of 100 million seconds). Only 50 unstable dwarfs can be found on the earth at most and most of them are descendants of the long-lived or were born by those long-lived dwarfs as a result of changes caused by high-energy ray attack from the universe; the rest a few are created artificially in laboratories or founded in nuclear reactors or nuclear explosions. How large is the map of dwarf

Z. Gan (✉) · Z. Zhang · J. Liang
Institute of Modern Physics, Chinese Academy of Sciences, Lanzhou, Gansu, China
e-mail: oamil@tom.com

© Zhejiang Education Publishing House 2021
G. Jin and G. Xiao (eds.), *Probing into the World of Nuclei*, China's Big
Science Facilities, https://doi.org/10.1007/978-981-16-0715-8_3

Fig. 1 The chart of nuclides

world exactly? According to estimates from scientists, there are about 7,000 dwarf inhabitants (Fig. 1).

Since ancient times, human beings have lived with those long-lived dwarfs and used them to make all kinds of tools and coins. About 120 to 130 years ago, scientists spent 20 years finding out these dwarfs one by one and then gave each of them a name (Fig. 2).

However, their ancestors are either ^{238}U, ^{235}U, or ^{232}Th. They produce one α dwarf to the next generation which also repeat the process. The dwarfs keep reproducing from one generation to another. Life cycle of each generation is long or short. Some dwarfs do not produce α dwarfs during the process, but release one electron instead to produce a member of another family which yet still produces the α dwarf. The process will continue until the lead family ultimately comes about.

2 New-Born Dwarfs

The "training base" of the dwarf combat force develops very fast, along with larger scale and better equipment. Under this condition, not only can the thin warriors such as protons and helium be trained to become qualified fighters, but also can the fat warriors, such as the calcium (Ca)-48 dwarf with 20 protons and 28 neutrons in the "belly", and the uranium-238 dwarf be trained on a larger scale to race against the light. Such a competent force can fight against various "castle owners" and create

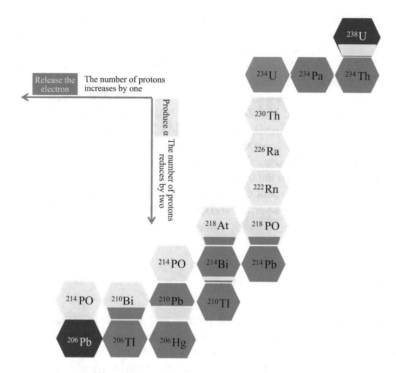

Fig. 2 Genealogy of the ^{238}U family

many new dwarfs. During the past 60 to 70 years, new dwarfs have been continuously registered after countless battles, combinations of two (fusion) or divisions of one into two (fission), secret migrations (transfer), multiple births (fragmentation) etc. Total number of the dwarfs has been increased to 3,473 as of now and most of them inhabit in subdivisions of the U family.

In the known world of dwarfs, there are 118 families and 3,743 inhabitants. Some of them like to stay peaceful instead of being naughty; some are very energetic and likely to change their appearance all of a sudden without being noticed; some are so skinny that a gentle wind may blow them down; some are so fat that the fat seems to be falling down! Since there are so many inhabitants, how do they manage to settle down in this dwarf world?

A look at the history reveals that the first inhabitant was registered by Rutherford in 1908. He put the matter that could release α into the electromagnetic field, measured how far the particle deviated from the original path, and calculated the amount of its charge—equivalent to that of two electrons (2e), and its relative mass—3.84. Based on these results, he believed that the α particle is the Helium nucleus (two protons and two neutrons). He then weighed each dwarf with a special approach for about 30 years and ultimately decided their proton number. This helped almost all stable dwarfs find their homes.

How do the rest 3,000 households settle down? Remember the battle among dwarfs? The battle is a good way to register new dwarfs. Over 100 years ago, Mr. and Mrs. Joliot Curie were the first to adopt the battle to register several new dwarfs. In the ^{238}U genealogy, there is a dwarf named ^{210}Po. This dwarf is very active and has a life expectancy of about 200 days. It also produces a small α particle from time to time and becomes a stable ^{204}Pb as a result. This little α particle was born fast and used by Mr. and Mrs. Joliot Curie to battle against boron (^{10}B, 5 protons), magnesium (^{24}Mg, 12 protons), aluminum (^{27}Al, 13 protons)—the "castle owners". They also used a Geiger counter to measure rays emitted by new dwarfs produced after the collision to calculate their life cycles and found that the new dwarfs produced after the "battle" between the α particle and the "castle owners" have different life cycles. They carried out careful study and found that the difference occurred after the original owner ate the α particle and spit out neutrons or protons which gave birth to ^{13}N (7 protons), ^{28}Al, ^{30}P (15 protons).

3   New Continent Discovered

Originally, scientists thought that the protons and neutrons in the dwarf's "belly" were randomly mixed together and looked like a drop water. After calculation, they believed the fattest dwarf had a maximum of 104 protons. Yet later research found that this is not the case. Protons and neutrons are actually layered like electrons in atoms. In this case, the fattest dwarf has more protons, especially that dwarf with 114 protons and 184 neutrons, and enjoys a very long life. The dwarfs centered on the fat dwarf are also very fat and long-lived. They form a so-called overweight area (island of stability of superheavy elements) where the number of protons in each dwarf's belly is more than 104 and the number of neutrons is above 150. Everyone wants to explore and develop this area of undiscovered sceneries, which requires tremendous hard work. Let's see what obstacles or difficulties lie in the journey to a fat dwarf that weighs 118 for example.

(1)   Calling up the Calcium-48 Dwarf Force

To find the fat dwarf that weighs 118, dwarf force, we need to first know about the calcium-48 dwarf force.

Footprints of the calcium dwarf family are ubiquitous on the earth—soil, water, plants animals etc. The Ca family has 20 protons for which reason they rank the 20th in the periodic table. As mentioned earlier, numbers among the dwarfs such as 20 and 8, 28, 50, 82, 126, and 184 are some magic. When the protons or neutrons in the dwarf's belly number as such, the dwarf appears to be more stable. Among the dwarfs discovered so far, 21 are registered inhabitants in the Ca family; most of these inhabitants have a short life cycle ranging from 162.7 days to 1 billionth of a second; only 6 are very stable. The six members have 20, 22, 23, 24, 26, and 28 neutrons in their belly. The inhabitant with 20 neutrons who account for two "20" magic numbers is among the vast majority of the Ca family which account for 97% of the member.

For the Ca family, 28 neutrons are really too many; yet 28 is a magic number and inhabitants with this number of neutrons and 20 protons are very peaceful. But these inhabitants are not so many, only 0.187% of the total. To find the No. 118 overweight dwarf needs an army force made up of the 0.187% Ca-48 dwarfs. How to separate this minority from other family members?

Let's first understand the magnet. If you have played with a magnet, you would know that it has two poles: N and S. When N and N or S and S approach each other, there will be repulsion; when close, N and S will attract each other. If N pole of one magnet faces S pole of another magnet with a certain distance kept, the field between and the surrounding area will be full of magnetic force. We call such a space a magnetic field. Let us look at the electric current, which is the flow of charged particles. When the charged dwarf passes through the magnetic field (N is down, S is up), it will be pushed by the magnetic force. The positively charged is pushed to the right side, and the negatively charged is pushed to the left. The more charges the particles carry, the stronger the magnetic force and push will be. When the electric and magnetic forces equal, the lighter dwarf (with smaller mass) will be pushed further.

In this way, after they pass through the magnetic field at the same speed, dwarfs with different charges will arrive at different locations. This approach can be used to separate the Ca-48 dwarfs. First, all members in the Ca family are heated in a small pot, so that each dwarf loses one electron on the outer wall of their fort. Meanwhile, voltage is applied to the pot to attracting the electrons and push them to quickly pass through the magnetic field. In this way, various twins of the family are separated. The Ca-40 brothers are the lightest and thus pushed farthest; the Ca-48 is the heaviest and thus pushed away in smallest distance. Between the two, there are also Ca-42, Ca-43, Ca-44, Ca-46. Fortunately, Ca-46 twins take up just a small proportion and allow for sound separation of the Ca-44 and Ca-48 twins. The machine that separates different members in the same family by the force of the magnetic field is called a magnetic separator. It takes a long time to get enough pure Ca-48 brothers. About 300,000 h, or 34 years are needed to obtain 1 g of Ca-48 (about 1.25 trillion trillion members) with the magnetic field which allows for 11,700 billion one-charge Ca-48 s to pass through per second and gather together. One year will still be needed even if 34 such machines work at the same time. It is also feasible to select Ca-48 with chemical approaches, which nonetheless also takes a considerable amount of time (Figs. 3 and 4).

With all Ca-48 members available, trainings are then required to build a qualified troop. The entire training process is quite complicated. Initially, all members are hiding in their own forts instead of in a unified team. The first step therefore is to place all Ca-48 castle owners along with their individual forts together in a high-temperature furnace, destroy the wall of each castle with powerful sand guns and remove as many electrons as possible from the wall. Second, apply a high voltage to the furnace exit to attract the Ca-48 castle owners along with a few remaining electrons on the wall together and make them pass through the magnetic field. Finally, choose owners of the same "clean" level (with same charge) and send them to the training camp.

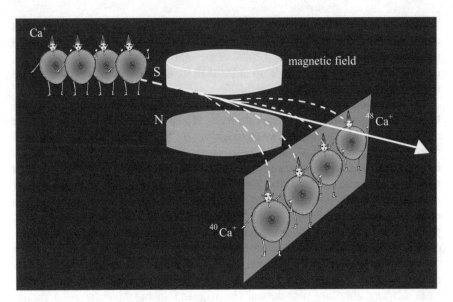

Fig. 3 Dwarfs in the magnetic field

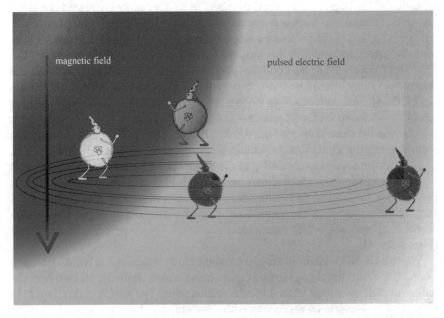

Fig. 4 The Ca-48 force are spiraling at an increasingly faster speed in training

The Ca-48 castle owners are constantly running in the multi-track magnetic field training camp. Each lap offers them charge at a fixed location from the applied voltage (some training camps have two charge stations for each lap). The dwarfs run increasingly faster as charges are applied. Then at one point of the track, they jump to the adjacent outer ring track. The process goes on and on; the faster the dwarfs run, the bigger their lap will be. After a few hundred laps, they reach the exit of the training camp. The exit is very narrow, and only those who have reached the required speed can run out of the exit and participate in the battle. About 5–6 trillion Ca-48 castle owners come out of the camp at qualified speed per second and form a mighty force that surges forward along the planned route to its target. The force will also be adjusted on their way forward so that they can gather together from time to time and will not get hit somewhere near the route to avoid unnecessary sacrifice.

(2)   The ^{249}Cf Target under Construction

Before constructing the ^{249}Cf (Californium, 98 protons, 151 neutrons) target, we need to first know how ^{249}Cf comes about. The heaviest stable element on the earth is uranium (in the symbol U with 92 protons; rank the 92nd in the periodic table). The U family has three relatively stable members, namely ^{234}U, ^{235}U, and ^{238}U. Among them, ^{238}U is in an absolute dominant leader, while ^{235}U only accounts for around 0.7% of the total and ^{234}U is even less, almost negligible. Although the twins of ^{235}U take up a small percentage, their role is appreciable. The ^{235}U dwarfs love to eat slow-moving neutrons. Unfortunately, the eating brings to them a serious fever which makes them spit out two or three neutrons and split into two at the mean time. Then the ^{235}U dwarfs turn into members of two other families with relatively fewer protons. They fly to their new home like a high-speed rocket in opposite direction, passing through many forts and finally settle down with enormous heat released simultaneously. The neutrons that are spit out hit around in various forts, run gradually slower and may be eaten by other ^{235}U members. If this process is to happen continuously with more neutrons spit out, more ^{235}Us need to be brought together. If the proportion of ^{235}U in the U family is increased to 5%, the U family can be used as fuel for nuclear power plants to generate electricity. If the ^{235}U takes up a dominant proportion in the U family, such as more than 95%, the U family can be used as fuel for neutron furnaces or to make nuclear weapons.

^{238}U also eats neutrons. If it eats a neutron that runs very fast, it will become a member of the other two families like the ^{235}U. If it eats a very slow neutron, it will have higher body temperature, spit out a sand (electron), emit a few invisible lights, and finally settle down. The proton in its belly will increase by one; its position in the periodic table will be one step back. ^{238}U ultimately turns into ^{239}Np (Neptunium) with 93 protons.

Fig. 5 Special neutron
furnace used to create the
^{249}Cf dwarfs

Fig. 6 Separation of the
forts in Cf family

Similarly, Np can also eat neutrons, spit electrons and become Pu (Plutonium, 94
protons); Pu also eats neutrons, spits out electrons and turns into Am (Americium,
95 protons) (Figs. 5 and 6).

By repeating the neutron-in-electron-out process, it can then produce heavier Cm
(Curium, 96 protons) and Bk (Berkelium, 97 protons) until members of the Cf family
come into being. Cf is the 98th element in the periodic table.

In order to gain ^{249}Cf, we need to first of all make a special furnace which produces a large number of neutrons in a short time. This kind of furnace is called high-flux neutron reactor. Unlike reactors in nuclear power plants, this furnace is small and its fuel is ^{235}U—the dominant leader of U family. In this furnace, more than 2,500 trillion neutrons pass through the fingernail-sized area every second. With the furnace is ready, put the prepared mixture of Am and Cm (targets) at the center first as neutrons are denser. Let the castle owners in the targets eat as many neutrons as possible. After about 24 days when the furnace consumes a lot of ^{238}U fuel, the number of neutrons drops and the fuel needs to be replaced so that the target castle owners can eat more neutrons. After five or six rounds, it takes about a year to take the targets out and transport them to special chemical plants to separate the needed owners, including Bk, Cf, Es and Fm (Fermium, 100 protons). The remaining owners who have not changed much are also very precious. So don't throw them away, just send them back to the furnace and let them eat more neutrons.

Of course, among the separated families, the majority is the Bk families, followed by the Cf families; both the Es and Fm families are few. In the Cf families, there are ^{249}Cf, ^{250}Cf, ^{251}Cf, ^{252}Cf among others. The ^{249}Cf has the largest number of members. When all ^{249}Cf members gather together, their size is only that of a sesame seed, about a few thousandths of a gram. To pick out the ^{249}Cf members, we need to filter over and over again to separate other members from their family and make the ^{249}Cf members account for over 98% of the total. They are indeed priceless treasure painstakingly obtained.

Next, they are placed in nitric acid to swim and find a nitric acid partner. Together with the partner, they are then put in a special plating tank. With electric currents, castle owners of the Cf family are firmly fixed on the 1.5-micron thick titanium foil and become the target for synthesis of the overweight No. 118 element. The Cf castle owners are very lively and soon mingle with the oxygen castle owners and create gray oxides similar to rust. The result is that the target does not look as bright as a stainless steel spoon.

(3)   Formation and Identification of the Overweight No. 118

With the ^{48}Ca force and the target for offensive ready, the first step is completed. Next comes the real battle. The target needs to be cut into several small pieces and then attached to the outer edge of a rotatable wheel to form a ring. When two trillion of ^{48}Ca take the offensive, temperature of the target will rise very high. If a place is continuously attacked, it will not be long before the target is smashed without any traces found. Therefore, when ^{48}Ca combatants start attacking the target fort, the wheel attached with the target has to run constantly.

Many castle owners will be born in the battle, and they will be surrounded closely by the combatants to pass through the target. In order to separate them to find the overweight, the new owners need to be introduced to a large iron cabinet called separator. Powerful magnetic fields are applied above the cabinet which is also home to lots of hydrogen forts—probably 20,000 trillion per cubic centimeter. A large number of combatants who do not participate in the actual battle and those who get hurt little are soon driven out of the original path by the magnetic field to a designated station. The reason is that they run at a similar speed of the combating force and weigh not much, coupled with absence of "clothes" (few electrons follow them on the periphery of the fort).

The rest are new and slow-walking overweight family members. About several hundreds of them are born each second. They continuously collide with the hydrogen forts in the magnetic field and seize the opportunity to pick up electrons from the forts to repair their destroyed walls. Gradually, they slow down and don't have enough strength to take electrons from the forts. As a result, there is usually an electronic gap on their castle walls. Under this situation, they cannot but slowly proceed in the magnetic field according to the planned path and finally gather together in front of a small exit channel for identification.

How to decide whether these heavy castle owners are the overweight we are looking for? Well, this is very complicated. For this purpose, several passes are set: the first is the speed at which they walk, which requires accurate timing at the start and end points; the second is where they stop, and this requires a precise locator; the third—the most important one—is whether or not their descendants match the known owners. Only with these passes can we calculate the number of protons and neutrons in their original belly.

The key to the first pass is the accuracy of the timer. Although the overweight runs relatively slow, they can still finish hundreds of kilometers per second, much faster than the rocket. Currently, the timing accuracy has reached 10 billionth of a second which is sufficient to meet the requirements. The key to the second pass is precise location, which is also not a problem. Positioning accuracy of the current detector has reached several micrometers, which is much more precise than the GPS. The third pass is the key to success. To this end, five cubic detectors of 4 cm length and high positioning accuracy are used to make a barrel that faces the direction in which the overweight flies to. The five detectors can be used to not only accurately locate, but also accurately detect the running speed or kinetic energy of the newcomer and its new helium princess (Fig. 7).

Generally speaking, the overweight members created currently are too fat, too heavy, and unstable. Most of them are born from helium princesses who produce certain kinetic energy. The members also keep advancing in the periodic table of elements from generation to generation till they find stable known owners.

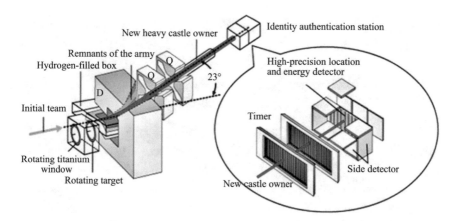

Fig. 7 Diagram of separator structure

In order to reason on the basis of the known owners which overweight family the members' ancestors belong to, we need to record every helium princess's birth time, location and energy. The result is that owners of the new castles formed with considerable hard work can only live for less than one thousandth of a second. The first generation of the helium princess produce kinetic energy at 11.65 meV, making her two steps forward in the periodic table; after another one hundredth of a second, another helium princess is created and its kinetic energy is 10.80 meV, two more steps forward in the table; the owner of this position will split in two (about 70% of the chance) after 0.6 s; sometimes it gives birth to a helium princess (approximately 30% of the chance) whose energy stands at 10.80 meV. In this case, the owner takes two steps forward.

Please note that the life cycle and the energy of the helium princess the owner gives birth to basically match the data of an owner with 172 neutrons in No. 114 position. This indicates that this owner should be the No. 114 castle owner with 172 neutrons. Then, after the No. 114 owner gives birth to the helium princess and splits in two new owners whose life cycle is 1.9 thousandth of a second, similar data are found in an owner with 170 neutrons in No. 112 position. Therefore, the last two owners are identified: one is the No. 114 castle owner with 172 neutrons and the other is the No. 112 castle owner with 170 neutrons. Based on this information, we can reason that No. 118 position can be reached either six steps backwards from No. 112 position or four steps backwards from No. 114 position which are consistent with the total number of protons in the ^{48}Ca and the ^{249}Cf castle owners. This information conforms that the overweight members' ancestor is one member of No. 118 family who has 176 neutrons in the belly. That is to say, scientists have created a castle owner in No. 118 position that has 176 neutrons.

Later on, this family is called Oganesium by the international community. Its Chinese name is 𰾢 (Fig. 8).

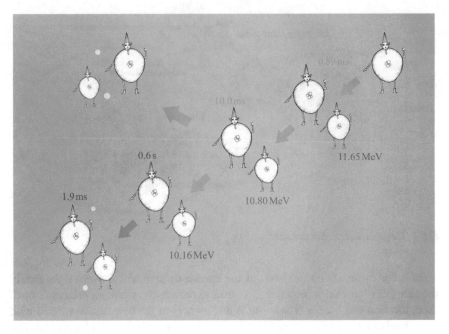

Fig. 8 Transformation of the atomic castle

Oganesium is named after a Russian scientist who has long researched the overweight dwarf to honor his significant contribution. This new overweight dwarf is the result of cooperation between Russia and the US. Russia provided the ^{48}Ca force (including training base—accelerator) and the identification device for No. 118 family; the US provided ^{249}Cf target. Together, the two countries registered new inhabitant in the farthest No. 118 position.

Up till now, thanks to hard work of scientists in the past 70 years, different brothers of 26 owners from No. 93 position to No. 118 position have been created. The new members expand to a large degree the map of the dwarf's world. This means that a dream of scientists has been realized.

4 Future Prospects of the Dwarf World

Does the world of dwarfs have any space for more inhabitants to be registered? The answer is definitely yes. Many years ago, scientists theoretically calculated the area of the world and reckoned that there was space for about 7,000 dwarfs to be registered. That is to say, there are 3,000 dwarfs are waiting for registration. What is address of the waiting dwarfs for registration? Scientist set rough boundaries for that. From this we can see that a lot of open space on the right side of the world map is available. Of course, this is theoretical calculation. In new places, will the dwarfs

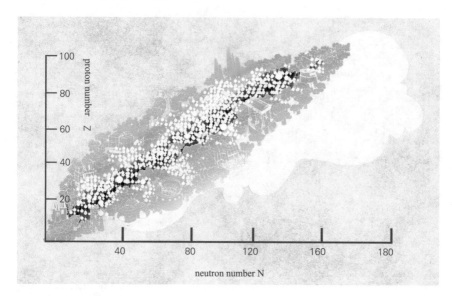

Fig. 9 The map of the dwarf in estimation

abide by the existing rule? For example, will 20, 50, 82, 126, and 184 still be magic numbers? To what degree will the dwarfs be overweight? In particular, where is the center of the charming overweight island? Answers for these questions require more hard work from scientists. If you want to join this expedition and enjoy the charming sceneries, try to accumulate as much knowledge as possible (Fig. 9).

Check How Much the Dwarf Weighs

Meng Wang, Zhengguo Hu, and Mingze Sun

Abstract Can we weigh the dwarf with the world's most precise scale? Definitely not. Then, how do you weigh the dwarf, and make more accurate results?

1 Never Short of Tools to Weigh the Dwarf

Can we weigh the dwarf with the world's most precise scale? Definitely not. Then, how do you weigh the dwarf, and you need to be very precise?

One day, the dwarf "Fatty Uranium" sat quietly in his castle, thinking, with his chin on his hands: "I know I'm fat, but how much do I weigh exactly?"

Well, who doesn't want to know about his weight? But the dwarf is too small to weigh on an electronic scale. Then, how can we weigh someone with such a small volume as the dwarf?

In China, everyone knows the story of "Cao Chong Weighing the Elephant". Cao Chong summed up the weight of the stones in the boat and got the weight of the elephant based on the principle of buoyancy. This is an indirect method of measurement. Can we also weigh the dwarf with an indirect approach? Yes, of course!

(1) Siblings of the Dwarf—The Discovery of Isotopes

In the world of dwarfs, everyone is very "light" in weight. If we use "gram" as the unit of mass, it is very difficult to describe their weight accurately. Therefore, scientists define the unified atomic mass unit (symbol: u) as one twelfth of the mass of a neutral atom of ^{12}C, or we say the mass of the owner of the castle of ^{12}C together with the electrons on its wall, i.e., 1 u $= 1.66053886 \times 10^{-24}$ g (As the mass and energy are interchangeable, we also define "u" as 1 u $= 931.496$ MeV). To measure an object with such a small mass, a very precise scale is needed.

M. Wang (✉) · Z. Hu · M. Sun
Institute of Modern Physics, Chinese Academy of Science, Lanzhou, Gansu, China
e-mail: oamil@tom.com

© Zhejiang Education Publishing House 2021
G. Jin and G. Xiao (eds.), *Probing into the World of Nuclei*, China's Big
Science Facilities, https://doi.org/10.1007/978-981-16-0715-8_4

In 1913, a British scientist named Joseph John Thomson invented a method to determine the weight of the dwarf by measuring the deflection of charged particles in electric and magnetic fields. First, dig out some electrons on the walls of the dwarf's castle, or simply drive the dwarf out naked and send them to the electric and magnetic fields, where the dwarf will deviate from the original track and turn involuntarily to a certain direction. When the speed of the dwarf and the intensity of the electric and magnetic fields are fixed, the deviation is only related to mass-to-charge ratio of the dwarf. And then, the weight of the dwarf can be calculated according to the deviation distance. Thomson used photographic plates to show the footprints of different dwarfs. He was surprised to find that the neon (Ne) dwarf walked two different paths. Thomson counted and counted, and finally figured out that there are two brothers in Ne (isotopes)—^{20}Ne with 10 neutrons and ^{22}Ne with 12 neutrons. Later, Thomson's assistant Francis William Aston invented the mass spectrometer that employs electromagnetic focusing, and found that each family consists of different brothers, with no less than 212 dwarf brothers in total (Fig. 1).

This is a direct method of measurement. However, such method did not allow accurate mass determination. Measurement with this method is accurate to a few tenths, which means, it only serves to separate different brothers in the same family. The dwarf also wants to know whether there is any difference between his weight and the total weight of neutrons and protons in his stomach, because it is directly related to the length of his life. In addition, the distance between neutrons and protons in the stomach of a dwarf also causes a slight difference in weight, which determines whether the dwarf can survive. And apart from all these, how does the sun shine all the time? Where does its energy come from? How long does the light during the supernova burst last? To find these answers, we must identify the weight of many dwarfs, accurate to a few hundred billionths. So is there a perfect way to accurately measure the weight of the dwarf.

(2)   Indoor Sprint

The stainless steel tube not far away from the training fields of the dwarf troops is their sprint racing track. On the track, other dwarf castles are mostly cleared, and two high precision intelligent timers are set up at both ends. Take dwarf oxygen as an example. When dwarf oxygen dashes out of the training field, the departure time is recorded at the starting point of the runway by the first intelligent timer; when it arrives and hits the baffle of the second timer, the timer not only records the arrival time, but also information on its energy. With these data, we can easily calculate the weight of the dwarf! The dwarfs have run 50.0155 m in 1.2011×10^{-6} s, and the energy is 143.75 MeV. As the energy equals to half the product of the square of velocity and body weight, we can get the body weight of the dwarf oxygen, which is 14899.2 MeV (2.6560×10^{-26} kg). We may as well extend the length of the track to improve the accuracy of both distance and time measurement. Nowadays, in a 100 m race it's easy to be accurate to 1/1000 s. For the dwarfs' race, this can be accurate to a few hundred billionths of a second, but the distance is measured in microns, or 10^{-6} m. If the track is too long, the accuracy of distance measurement would be

Fig. 1 Mass spectrum from Thomson's 1913 publication (different markings on the photographic plate represent different isotopes)

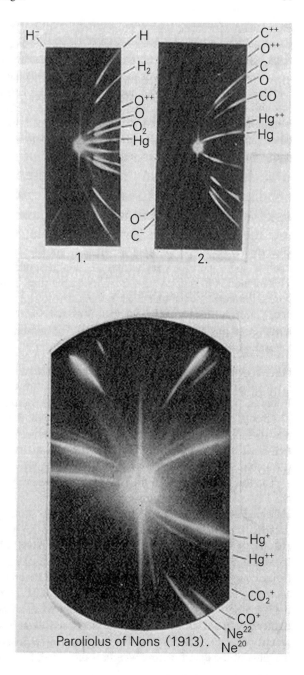

even lower. Moreover, energy is measured at 10^{-4}eV, which determines the overall accuracy of measurement. For instance, in 1982, a French scientist lengthened the track distance to 82 m, and obtained the speed value by measuring the running time, thus identifying the weight of the dwarf. Yet the precision of his result was less than hundreds of thousands of, but not only 100,000. Of course, the distance of the track can also be increased to further improve the measurement accuracy of the dwarf's weight. But limited by space, it cannot be extended indefinitely. More importantly, due to limitations in the accuracy of energy measurement, the overall measurement precision with this method is greatly restricted.

2 Newcomers Weighed in the Race

Over twenty years ago, German scientists invented a "storage ring mass measurement method" to measure the weight (mass) of hundreds of dwarfs with a higher precision. This method is to keep the dwarf running on a ring-shaped track, measure the time it used for each lap, and calculate its weight based on other data obtained. Ten years ago, China also measured the weight of the dwarf with the cooling storage ring. The accuracy of this method may reach one in ten million. Let's see how they measure the weight of the dwarf.

(1)   Fast and Slow—Isochronous Mass Spectrometer

In the second chapter, we have visited the large training field of the dwarf troops. The troops coming out from the training field fought with the castle owner, and many new dwarfs were born in the fight, all running out of town naked at the same speed as the original troop. They continued to run forward and entered a new large stadium (the cooling storage ring) through a passage (Fig. 2).

The track measures 128 m. This arena has a new rule, i.e. classify new fighters and those who are equal in charge and weight ratio (called charge-to-mass ratio) are classified as one type. The fighting dwarfs of the same type need to ensure that the time for running each lap is the same. To achieve this goal, those who run faster take the outer side of the track and those who are not physically strong enough to run fast take the inner side. In this way, the time for each dwarf to finish one lap only depends on the charge-to-mass ratio. All dwarf fighters in the arena come in and keep running according to the rule. Now it's time to measure. The time of each dwarf to finish one lap is measured first. With this information, we can know what type of dwarfs each fighter belongs to. In each type, dwarfs whose weight is known are found out first to identify charge-to-mass ratio of their types. On the basis of this data, we can decide the mass of the dwarf, whose weight is unknown, based on their charge.

(2)   Bright Eyes—High-Sensitivity Detector

When dwarfs are running on a ring track, we need an excellent referee to judge among the dwarf siblings, who run fast? How fast are they? Who are slow? How slow are

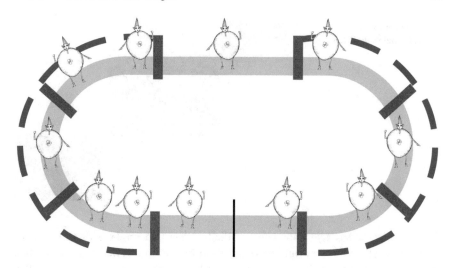

Fig. 2 Dwarfs on the "track"

they? Only with this information can we find out the different weight of each dwarf sibling.

Dwarf fighters in this track and field competition have received trainings in field and gained strong running capacity. Their speed has reached 2,00,000 km per second which is two thirds of the light speed and over 15,000 times the speed of a rocket carrying a satellite into the sky. Therefore, if the time of the dwarf for each lap is to be recorded accurately, we need a referee with bright eyes and a very sensitive timer. Who can take this role? The answer is "high-sensitivity time-of-flight (TOF) detector" (Fig. 3).

Fig. 3 Diagram of high-sensitivity TOF detector

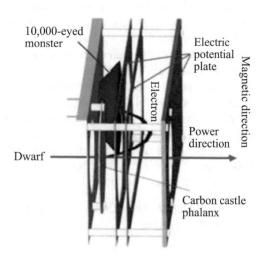

Fig. 4 Diagram of the microchannel plate (MCP) structure

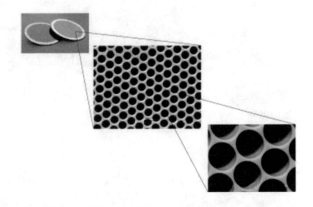

This detector has two superpowers. First, it has two bright eyes to catch the moment in which the fast-running dwarf flash by. Second, it has extremely quick response to the moment in which the dwarfs flash over and accurately record that moment.

Why does this device have such strong power? To answer this question, we need to know about its structure in the first place. It has a carbon castle phalanx (carbon film) of only one micron, three charged plates (electric potential plate), one two-pole magnet and a glass sheet of 0.2 mm thick and 40 mm wide (microchannel plate). This glass sheet is not an ordinary one in that it is made up of many small tubes with 5 microns thick. Let's look at the usage of each component (Fig. 4).

The carbon castle phalanx is a marking line of the ring track. Each time they run a lap, the dwarfs will cross the line once and release some electrons as a result of collision with the fort. The referee then uses the electrons to record the crossing time.

The electric potential plate is arranged in parallel with an external secondary magnet, three potential plates are arranged in parallel. Different voltages are applied to form an electric field. The external magnetic field of the secondary magnet is perpendicular to the electric field. In this case, be it the inside or outside of the ring track, the time for the released electrons to pass through the marking line and get caught by the eye of the referee is the same.

MCP is the eye of the referee. It receives and amplifies the electrons with which the referee takes records. About ten million little bright eyes are on this board, making it much more powerful than the thousand-eyed monster in *the Journey to the West*.

With such device in place, the tasks will surely be completed brilliantly. Scientists in the Institute of Modern Physics accurately have weighed multiple dwarfs that can only live for a few thousandths of seconds. The dwarfs are: ^{41}Ti, ^{45}Cr, ^{49}Fe, ^{53}Ni, ^{63}Ge, ^{65}As, ^{67}Se, ^{71}Kr… their homes are on the left edge of their continent and neutrons in their belly are much less than protons.

(3)   Surveying the Sky—the Relationship Between Mass Measurement and Supernova Explosion

When you look up to the sky at night, the stars appear to twinkle. Do you know how this happens?

Astronomers tell us that only stars shine by themselves and shine steadily for a long time, just like the sun. Although the moon looks bright at night, the light is not its own, but the light reflected by the sun. Apart from the self-luminous stars, there are also explosive luminous stars in the sky. This phenomenon is called supernova explosion. China observed such luminous stars in 1054 of the Song Dynasty, being the first to achieve so in the world. On the first 23 days, even during the day, these stars are as bright as Taibai Jinxing (or Venus). The explosion gives out light which takes two years to dim. Remains of the explosion are Crab Nebula.

The light visible to our naked eye is called visible light. There is also and a variety of invisible light, such as X-ray, γ ray, a variety of dwarfs on or outside of earth and smaller particles.

Be it light of stars or supernova explosion, they are all powered by extremely fast collisions of various internal dwarfs driven by gravitation. Species of dwarfs in long-live stars are few; most of them use hydrogen and helium as raw materials and produce members that range from the lithium family to the iron family.

How did the dwarf family heavier than the iron come into being? To answer this question, we need to first of all know about celestial events such as the supernova explosion. Do you know? There is a star in the space called neutron star which is composed of neutrons and a few other particles. This star is not that big; its radius is somewhere between 10–20 km. Yet it's weight is huge, 1.4–2 folds that of the sun. New dwarf families come about when two neutron stars come together or the neutron star uses its magic power to absorb a nearby star of small mass into its body. These are also the cradle of many short-live dwarf families, especially overweight dwarf families. Some neutron stars give out extremely strong X-rays in a very short time; this phenomenon is called X-ray burst.

X-ray burst is an extremely fast process in which the neutron star (the host star) plays its magic power to absorb a nearby star of small mass into its body. In less than a second, X-ray brightness (intensity) can increase by 20–50 times; its total power amounts to 10^{32} watts. The continuous luminescence can last for more than ten seconds. The burst breaks out repetitively after a period of time; some bursts can repeat for several times, and effects of each time are different. This is something related to how the next neutron star will evolve, how the dwarf family originates in the universe and how much proportion each dwarf family accounts for. Currently, scientists are trying to reproduce the observed X-ray exposure curve. But it is also indispensable to acquire information on the life cycle and weight of each individual dwarf family member born in this process. The very slight difference in weight will result in a huge difference in calculation. For example, the ^{65}As dwarf gains weight less than 1/1000 of the original, but the duration of the calculated X-ray burst is almost 30 s shorter. In fact, the total luminous duration is less than 200 s (Fig. 5).

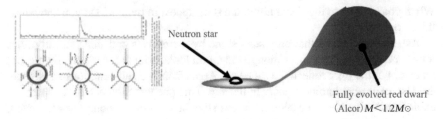

Neutron star

Fully evolved red dwarf
(Alcor) $M < 1.2 M \odot$

Fig. 5 X-ray burst diagram

3 Can You Weigh "Me"?

I'm neutrino, a very light-weight particle. I'm not charged. Some call me the "invisible man in the universe". I can run along with light and can pass through the earth almost without any obstruction. I have two brothers and our identities are interchangeable. We come from stars and supernovae explosion when some dwarfs spit out an electron and become members of other families. We are ubiquitous in the universe, 300 per cubic centimeter on average. Every second, hundreds of millions of particles like us pass through people's eyes. But no one knows our weight yet till now.

Scientists have invented an "ion trap". As indicated in the name, the idea is to sink ions (charged dwarfs) into a "pit". But this "pit" is a bit complicated with three deformations. One form of the "pit" features a slim-waist cylinder split to four equal valves. The trap also has a center both at the top and the bottom; both centers have a hole. The concave cap in the middle looks much like the antenna pan of a small satellite and sits on the slim-waist cylinder with an insulation mat in between. The upper and lower covers are applied with the voltage; and fast-changing voltages are also needed for the four valves of the cylinder. The cylinder with covers is placed vertically in a hugely strong uniform magnetic field. The slow-running dwarfs enter the trap from the bottom center and start rotating as if the are lost in the middle. Due to the strong magnetic field, the dwarfs rotate very fast, with a circle not that big though. In this case, voltage needs to be applied to two opposite valves on the cylinder to push the dwarfs outwards. Then two adjacent valves are applied with voltage in turn to make the dwarfs gradually rotate in an increasingly big circle with self-rotation at the same time. Apart from the newcomers, inside the cylinder are many other owners. Due to constant friction with the owners, the dwarfs gradually slow down. The big circle and the self-rotation circle are both getting smaller. As the dwarfs ultimately approach the central axis of the cylinder, the two circles overlap. Finally, a voltage is added outside the cylinder to push the dwarfs fly out from the hole at the center of the upper cover and hit the detector. When the voltage for bigger circle changes at the same speed of self-rotation of the dwarfs, the dwarfs will take the shortest time to fly to the detector. The weight of the dwarfs can more auxiliary be calculated with the strength of the magnetic field and through constant changes of that voltage which helps find out the speed for exit at the shortest time. (Fig. 6).

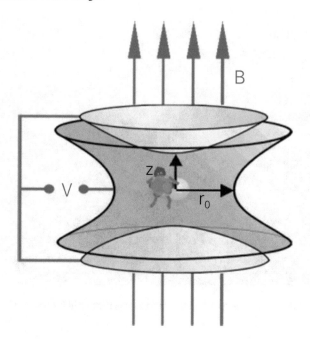

Fig. 6 Diagram of ion trap structure

This kind of ion trap has already weighed many dwarfs, with an error of one 100 billionth at most. However, the error in weighing neutrinos needs to be one trillionth. And for the weight of protium dwarfs, the error is less than two 10 billionths. Scientists at home and abroad began to make this kind of precision "ion trap" a few years ago. Although the device looks very big, its internal core is a tiny ion trap with a lot more supporting components (Fig. 7).

Fig. 7 Photo of the ion trap used for tritium neutrino experiments in Karlsruhe, Germany

Gallant Corps that Kill the Devil

Qiang Li, Zhengguo Hu, and Xuyi Huang

Cancer are the enemies of human health. Doctors have spared no effort to solve this problem. The dwarf corps can be the elites in fighting with cancer. Now let's learn how these dwarfs kill the cancer.

1 Excellent Corps

The aircraft flies at full speed in chaos as if it were wandering in the universe. After some time when I was almost falling asleep, a soft tone from the broadcast woke me up, "Dear passengers, as we are soon entering the skin tissue of humans, and there may be bumpiness ahead, please do not move and fasten your safety belt". The skin tissue is like a dense stardust compared with this aircraft which is no larger than the dwarf. While approaching the skin cells, the aircraft indeed underwent severe turbulence and eventually leveled out after coming through countless atom castles. "We are on this voyage to observe how the dwarf corps, the heavy ion beam, defeated the cancer giant corp." The aircraft broadcasted the aim of our voyage. It is destined to be a glamorous yet fierce battle.

The aircraft was floating in the deep of the skin tissue. The bizarre cell tissues around us seemed to be the distant galaxies. According to instructions, I took out my ultra high magnification telescope to observe the distant sceneries. What I saw were crowded atom castles which were mainly comprised of carbon (C), hydrogen (H) and oxygen (O) atoms. They were beating all the time and crowding against each other. I changed a new perspective just to see a bunch of dwarfs flying towards me. Looking through, I found that these dwarfs were carbon ions. They are well organized and flying rapidly towards the skin tissue, just like a fleet of aliens. It is said that "do

Q. Li · Z. Hu · X. Huang (✉)
Institute of Modern Physics, Chinese Academy of Sciences, Lanzhou, Gansu, China
e-mail: huangxy@impcas.ac.cn

© Zhejiang Education Publishing House 2021   69
G. Jin and G. Xiao (eds.), *Probing into the World of Nuclei*, China's Big
Science Facilities, https://doi.org/10.1007/978-981-16-0715-8_5

not judge a person by appearance." So do not underestimate these seemingly small dwarfs. Actually, they are the major force against cancers. Though they are generally small in size, they are quite gallant and well-disciplined in battles.

What kind of exceptional abilities do these dwarfs have?

(1)   Benevolent and Persevering

While combating the deep cancer tissues as ordered, the dwarfs of carbon ions try their best to protect the innocent normal cells and tissues along the way ahead. After a long journey, they finally arrive in the nest of the "enemies". Though they are so exhausted at the moment, they do not surrender. Instead, they burst all their energies and stimulate their comrade-in-arms, the electrons, to kill the enemies. This is the Bragg energy bees which are exclusive for the dwarfs and are used to kill the cancer cells effectively.

At the mention of cancer cells, everyone hates them. They are the cells that free from the management and control of the "Central Command". They reproduce in a crazy manner. They sack the nutrition from healthy tissues, erode healthy cells and eventually cause people to be slim, weak and dead.

The Achilles' heel of the tumor cell is the molecule of DNA. These DNA molecules are like fried dough twists spiraling into a "rope ladder". The side ropes are composed of numerous dwarf castles from different families arranged one by one in certain order, and the beams are built by dwarf castles from various families. These DNA molecules carry the generic instructions and codes, guiding the production of some organic compounds to ensure the endless reproduction and survival of tumor cells. Therefore, we can completely kill the tumor cells if we destroy these "ladders".

The flying carbon ions in high speed are like righteous and indignant warriors, destructing these "rope ladders" into pieces. Both the side ropes and the beams are cut off and never can be restored. As such, the tumor cells get a deadly strike while the surrounding healthy tissues are well protected.

Apart from carbon ions, the X-ray, γ-ray, and the proton are also major powers of anti-cancer campaign.

Usually, hospitals use X-rays to attack the tumor. The motto of the army of X-rays is "we would rather kill all for just one". So along the way, they may kill lots of innocent normal cells and get hurt seriously. When they get to the nest of the enemies, their fighting capacity has been so severely impaired that they could just cut one side rope of the ladder at best. Thus, DNA can soon repair the ladder. This rude way not only hurts the patients at the moment, but also poses danger for the days to come, because some normal cells may convert into tumor cells (Fig. 1).

(2)   Vary in Speed and Responsibilities

As we all know, the initial velocity determines the range of the bullet. Equally, the velocity of the dwarf corps determines how far they can get in human tissues. In order to destroy tumor nests of different depths, the dwarfs shall be trained to have

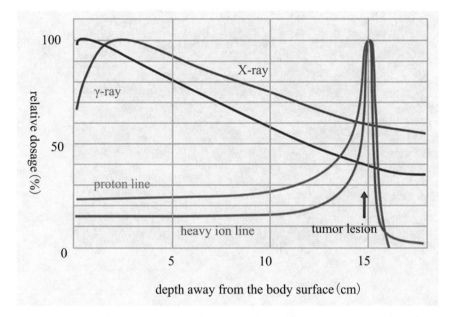

Fig. 1 Energy curves of various particles according to the changes of depth in human body

different speed or the ability to change their speed in the midway. The corps with low speed could deal with tumor in low places, and those with fast speed could confront tumor in deeper positions. If the tumors are in very deep places, dwarfs are required to work in unison so as to release energy and wipe out the enemies (Figs. 2 and 3).

(3)   Well-disciplined, Strict in Enforcement of Orders and Prohibitions

Dwarfs of carbon ions are corps that has undergone strict training. They are well disciplined and strict in enforcement of orders and prohibitions. Once receiving the orders, they are highly unified and quickly rushing into the tumor nest, ready for a bloody fight. Though corps of protons has similar explosive force with carbon ions, they are somewhat undisciplined and can easily get distracted and dropped out during the journey. And the position of energy burst is not concentrated. The corps of X-rays is even worse. They are undisciplined and disobedient, and often cannot tell friend from foe. Regardless of the numerous road signs and indicators along the way, there are still many unruly guys that leave without authorization, wander around and kill the innocent.

By contrast, the dwarfs of carbon ions are worthy of the title of the Bravest Corp. Given that they are highly disciplined warriors who never kill the normal cells by mistake, they are always dispatched to kill the tumors in very sensible parts of human body, such as the head and neck, near the large vessels in the liver, near the large vessels in the lung, and in the uterine neck (Fig. 4).

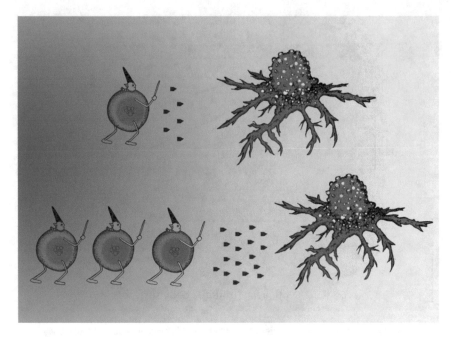

Fig. 2 The depth reached by carbon ions of different energy levels

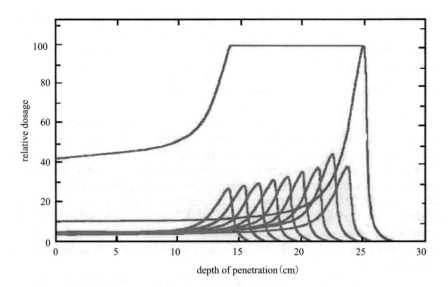

Fig. 3 Superposition of carbon ion beams of different energies in tumor target area

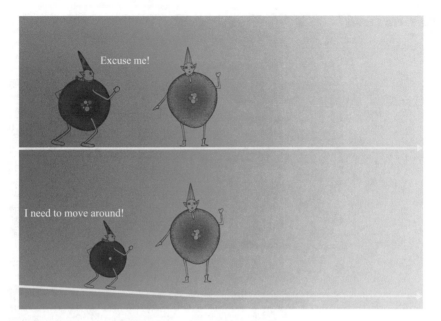

Fig. 4 Degree of concentration of proton lines and carbon ion lines along the beam line

As the dwarf warriors are electrically charged, they can be adjusted flexibly in their ways of attack. There are usually two forms, one is passive shape adaption, and the other is active shape adaption. In passive forms, some invisible hands like swing magnets, scatterers and multileaf collimators are put in their way to attack, and "mould" the corps into the same shape as the tumors. In active ones, the dwarfs first divide the nests of enemies into several blocks and each small block is set as one point of attack. And then they divide themselves into several groups. Each group adjusts their velocity according to the location of the blocks. Each group is responsible for one small block and guided by the scanning magnets, they destroy the enemies' nests. In a word, the ways of controlling and adjusting the corps are various, feasible, efficient and effective (Fig. 5).

(4)   Create New Dwarf Warriors which are Subject to Supervision

The track of dwarfs can be supervised online. Along the trudge, dwarf warriors can accidentally collide with the dwarfs in the surrounding cells. Some of their neutrons can be captured by other cells, making them short-lived. For instance, a C10 with 4 neutrons may not get slow. But once stopping, it will quickly give off positron. The positron meets the negatron and there emerges two γ-rays moving into opposite directions, which can be traced by Positron Emission Computed Tomography (PET) Camera. As these fragments are following the dwarf corps all the time, we just need to detect the signal sent by them and know the real-time position and number of the dwarfs. Such property cannot be found in regular X-ray corps.

Fig. 5 PET Camera

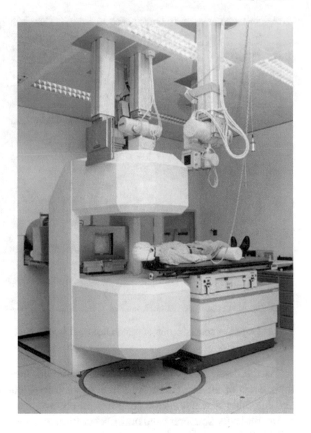

Up to now, have you updated your perception to the dwarf corps? Scientists just use this kind of property of dwarfs to deal with cancers in fixed-point blasting. In this way, the tumor cells are killed effectively and the surrounding tissues along the radiation channel are protected as much as possible. Therefore, the corps of heavy ion beam is entitled the most favorable rays in radiation oncology in the 21st century (Fig. 6).

2 Clear and Precise Targeting

The fight with tumor is a bloodless, smokeless yet perilous battle.

It is no coincidence that the dwarf corps of carbon ions assembles in established time and place and succeeds in killing tumors in proper manner. Along with the development of the times, killing tumor cells with armies is no longer a war of engagement but an information war based on massive intelligence. "Know me and know thee, lose none and win all." On the eve of the battle, by means of advanced detecting systems, such as CT, PET and PET-CT, information of the position, size,

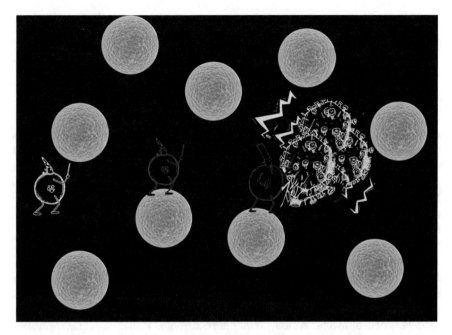

Fig. 6 The difference of heavy ion beams and X-rays and γ-rays in treating tumors

density of the tumor is well acquired. Based on such information, scientists make out targeted battle plan to kill the enemies and protect the healthy areas as much as possible. Here, the treatment plan is the General, who can command each dwarf to kill its enemies.

The CT and PET are the clairvoyants which intensively observe the circumstances in the tumor area, and send confidential information such as the army deployment, military configuration, and changes of the enemies to policy makers. Given such reliable data, policy makers can construct the 3D images of the tumor in a timely manner.

Image reconstruction is to copy the stereo structure of the enemies. With this real target, the command system is able to formulate a precise attacking plan and to ensure that in real cases the corps could kill the tumor cells.

The treatment plan system is the brain of the battle. As information technology develops, with its powerful arithmetic capability, the system could formulate several detailed war plans based on the information collected and attacking the established target. The plans include details like how to divide the battlefield, how to arrange the number of soldiers in each area, how to set the speed of the soldiers and how to deploy the military forces. Then based on the plan, a mock battle can be carried out and the plan which is closest to the desired effect will be selected after repeated verification.

It is the so called: "Once the target is set, corps are gathered to attack; battle plans are elaborately tested before adopted, to ensure the final success of tumor-killing."

3   Multiple Merits and Effects

How to measure an army's fighting force? Maybe the most obvious indicator is the comparative casualties. Killing three thousand enemies by losing two thousand soldiers is not cost-effective, which implies that the corps' comparative might is not prevailing. But killing three thousand enemies by losing three hundred is a good bargain. While fighting with tumors, the X-ray corps need to consume twice to three times the might of the carbon-ion dwarfs to kill half of the enemy. In medical science, this is called the "relative biological effect".

While corps of X-rays are in the battle, dwarfs of oxygen are like the boosters. The less dwarfs of oxygen ions, the worse the performance of the corps of X-rays. As its number increases, tumor cells swallow many dwarfs of oxygen ions, so among tumor cells there are scarcely any dwarfs of oxygen. On the one hand, if tumor cells increase slowly, there will be more oxygen ions. However, it is totally another case for dwarfs of carbon ions. As long as they are confronted with tumor cells, regardless the number of carbon ions, the carbon ions can always fight bravely. On the other hand, tumor cells will undergo several growth stages, including chrysalis, division stage and stable stage. In different stages, the resistant strength of tumor cells against the X-rays is different, and the fighting capacity of the dwarfs of X-rays is also in disparity. However, no matter in which stage, the dwarfs of carbon ions would attack the tumor cells in an all-out manner. Meanwhile, they would prevent cell cycle from happening. For instance, they can terminate or delay the process of cell division so as to finally kill tumor cells.

Given the two aspects above, the fighting capacity of the corp of carbon ions far surpasses that of X-rays.

4   Ready to Start

"The horn starts and the corps get assembled." The dwarfs of carbon ions have get ready to set foot on the journey to kill tumors.

Since 2012, Institute of Modern Physics of Chinese Academy of Sciences had started to establish demonstration set of heavy ions treatment for medical use in Wuwei. In 2015, it was completed and the beam was developed. This set combines a special circular accelerator with the synchroaccelerator. There are totally four treatment rooms: horizontal beam room, vertical beam room, horizontal & vertical beam room, and 45° beam room. The synchroaccelerator is 56.17 m, making it the world's smallest exclusive synchroaccelerator of heavy ion treatment.

Fig. 7 Model diagram of heavy ion accelerator for medical use

The set of Wuwei is China's first heavy ion accelerator for medical use with its own R&D and IP. Up to now, the set has done with the final adjust, comprehensive test and check and has gained the permit, ready to train the army of heavy ion dwarfs and bring blessings to all patients (Fig. 7).

There Is Much to Be Done in the Vast World

Libin Zhou, Genming Jin, and Jie Liu

How to make soybeans and wheat stronger and superior in quality? How to make flowers more colorful? How to ensure the normal operation of space flight? All these problems can be solved by armies of dwarfs!

1 Is Variation that Horrible?

(1) Basic Composition and Classification of Plants

I take the helicopter, fly through the microscope and come to the world of creatures. How marvelous the creatures are. Only creatures have the capability to transform disordered materials into ordered materials. So cute are those round balls squeezing together. Every cell is like a micro factory, where thousands of chemical reactions happen under precise control. This factory is operating in a cytoplasm whose shape is like a jelly. Inside the cytoplasm, there so many brothers and sisters who rely to each other: cell nucleus, endoplasmic reticulum, mitochondria and vacuole. Among them cell nucleus are the most important because it serve as the command center of this factory. How large is this factory of cells? Most are 10 to 30 microns, but the differences are quite tiny. The smallest independent cell is mycoplasma, with a diameter of 0.1 micron. The oocyte is a very big cell. For example, the oocyte of ostrich is as long as 25 cm, which is the biggest cell known.

There are two kinds of cell factories: plant factory and animal factory. Here we mainly focus on the plant factory. The outmost of it is a hard armor, the cytoderm. Without it, the factory may reduce to a green ruin. Inside the armor is a soft cytomembrane, which determines the border of the cell, produces and maintains the differential

L. Zhou · G. Jin (✉) · J. Liu
Institute of Modern Physics, Chinese Academy of Sciences, Lanzhou, Gansu, China
e-mail: jingm@impcas.ac.cn

© Zhejiang Education Publishing House 2021

G. Jin and G. Xiao (eds.), *Probing into the World of Nuclei*, China's Big
Science Facilities, https://doi.org/10.1007/978-981-16-0715-8_6

electricity environment, and controls the entry and exit of organic molecules. Without it, the normal operation of the factory would be in chaos. Outsiders and insiders cannot be distinguished. Organic molecules such as amino acid, protein, saccharides, nucleic acid are mixed together. It is just like a noisy fair. Unlike animal factories, plant factories can compound nourishment. The raw materials are quite simple: hydrone and carbon dioxide. There are the reaction towers called the chloroplast scattering in the factory.

As the sun shines, the chlorophyll and other pigments capture the energy in the sunlight and a series of complex photosynthesis is started. The hydrone divides into hydrogen atom and oxygen atom; the hydrogen atom combines with the molecule of carbon dioxide; oxygen molecule and glucose molecule emerge happily; oxygen molecule is given off into the air; and glucose is transmitted to other plants, providing energy or becoming starch and getting stored.

The shape of the plant factory is determined, to a large extent, by the armor of cytoderm. The cytoderm is a fairly orderly complexus, composed of various polysaccharide, proteins and aromatic compounds. Cellulose is the richest plant polysaccharide on the earth, whose basic unit is β-D-glucan. Cytoderms can be applied in wood, paper and textiles, and are also components of fresh fruits and vegetables, providing dietary fibre for human body. To improve the constituents of cytoderms is the target for human to develop industries of food processing, agriculture and biotechnology.

A creature cannot be formed by only one cell, with the exception of the monad. I come out from the microscope and get in front of the computer, ready to check some documents. I need to study the features of plant development. Cells need to split continuously to generate more new cells. Then after expanding and growing, they become the organs in plants, such as root, stem and leaf. During the growing stage, a single cell is not simply proliferating for its own sake; it exerts different influences on the organism. It is a kind of division of labor. Various types of cells in multicellular organism play various roles. Cell growth and cell division coordinates with each other, forming organs of certain shapes and functions, and then becoming a complete organism. And specialized cells and tissues are formed through certain locating or ways of sequencing. Cells can report to each other and exchange signal through small molecules and ions to control the destinies of cells. Meanwhile, the plant, as a whole, perceives and responds to the surrounding circumstance. For instance, sunflowers can trace the direction of the sun in the daylight.

One feature of plant growth is the forming of new organs at the peak. The overall character of plants is changing all the time. At the final stage while the embryo (baby plant) is going to come out, there are two kinds of cells, or called meristems, which generate the plant. The stem meristem generates the stem and the root meristem generates the root.

It is noticeable that a mature embryo of the mammal (baby animals) has the major parts of the body. If you pick a branch outdoors and plant it in your flowerpot at home, then with proper temperature and humidity, soon you will get an intact plant. The reason that plants can be generated from part of its body is that the destinies of plants are more changeable than animals, that is, in mature plant embryos, all cells are not fixed yet, and no special cells have been divided into germ lines. Given this feature, once using dwarfs of carbon ions to attack explanted plant tissues, branches or tubers, we can get new mutants in a short time due to the totipotency of plant cells. In this way, we create new varieties of plants and economic benefits.

I take the helicopter to go up to the sky, looking down to the earth. There are so many colorful plants. What on earth is the number of these plants? Actually, plants on the land are mainly spermatophytes, which can generate seeds. The existing spermatophytes are over 223,000 in five monophyletic groups: cycadales, gnetales, ginkgopsida, coniferales and angiosperm. Angiosperms refer to the spermatophytes which can bloom, including monocotyledons and dicotyledons. Rice, wheat, maize, sorghum, barley and oat are monocotyledons, and soybean and peanut are dicotyledons (Figs. 1 and 2).

(2) DNA, Chromosome and Gene

Now I take the helicopter, go through the electronic microscope and come to a smaller world of genes. As we all know, numerous creatures coexisting with us on the earth are determined by chemical "formulas" (genomes). Each "formula" exists in the form of chemical information, which are stored in spiral molecules called DNA and RNA. In living cells, hereditary information is stored in the form of double-stranded DNA. However, the genome of virus is various and somewhat special. It can be a single-stranded or double-stranded nucleic acid, a DNA, or a RNA. DNA and RNA are polymer molecules composed of nucleotide monomers, which are forming by 4 different nucleotide structural units.

Then what is the nucleotide? Each nucleotide consists of three chemical groups: base, pentose, and phosphate group. Bases include two kinds of purine bases and two kinds of pyrimidines. The DNA contains adenine (A), guanine (G), cytosine (C), and thymine (T). RNA also contains A, G, and C, but thymine (T) is replaced by uracil (U).

DNA was first discovered by Swiss biologist Johann Friedrich Miescher in 1869. Miescher proved through chemical experiments that DNA is acidic and rich in phosphorus. A single DNA molecule is quite large. The DNA double helix is so magical that no one knew its specific structure in the first half of the 20th century. Until 1953, Watson and Crick discovered that DNA is a double helix, which is a milestone in the studies of 20th century biology.

Fig. 1 Structure of plant cells

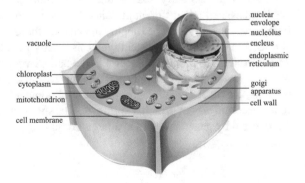

Fig. 2 Organs of plant

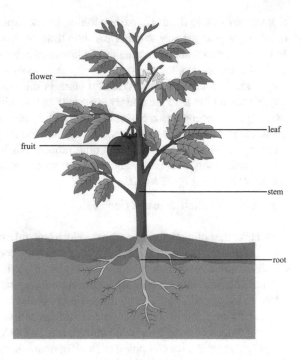

How does biological information pass from DNA sequences to biological functions? Through the research of ancient biologists, they usually carried out various wonderful genetic and biochemical experiments in the laboratory, and finally summed up the classic "central rule" of genetics, that is, genetic information passes through "DNA-RNA-protein" and eventually forms a complete organism.

In the nucleus of a cell, there are several DNA molecules of different lengths. Each one is called a chromosome, which has two copies, one from the mother and one from the father. The DNA in the chromosome is coiled itself and is entangled in other chemicals, such as proteins. The double helix structure of DNA is linked by bases, and the sequence of four bases in DNA constitutes a DNA blueprint of a distinctive cell. Each of the three base sequence forms a codon, and each codon corresponds to an amino acid sequence or a signal generating molecule. As such, a DNA molecule can carry a large amount of amino acid sequence information and can be translated into different proteins. A gene is actually an operational guide to a living organism, or it can be roughly understood that a gene is a specific region on a chromosome that controls protein synthesis. It tells the cell when to do something and directs proteins to properly function in the body. Everything proceeds in an orderly manner. However, there are only partial sequences in genes that can be translated into proteins, and other sequences just provide additional information, such as control the time, place and way of protein expression. The shape, stem, leaf and flower of a plant are affected by genetic factors to a large extent (Fig. 3).

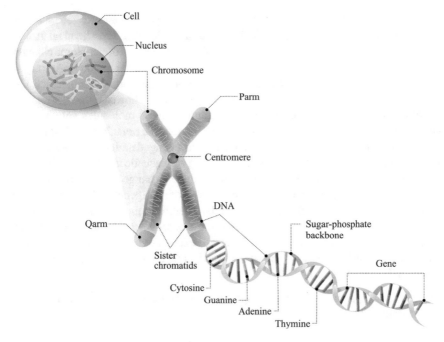

Fig. 3 Gene, DNA, chromosome and cell

(3)   Radiation and Functions of Materials

"So tired" was I after I came out of the magical world of plant and learned a lot of biological knowledge. Now, let's take a look at the physics of radiation. The invisible radiation is amazing. It can transmit information (with radio waves) in the blink of an eye; it can boil water (with microwave) in a few minutes; it can kill pathogenic microorganisms (with ultraviolet rays) and tumor cells (with X-ray radiotherapy), and it can also produce new crop varieties (mutation breeding) for farmers. These processes involve interactions with organisms and can be attributed to the mutual transfer of energy between radiation and matter. Radiation is usually divided into two categories: ionizing radiation and non-ionizing radiation. Ionizing radiation can ionize matter molecules, that is, the process of knocking out electrons of atoms to produce positive and negative ion pairs. By contrast, non-ionizing radiation can only cause changes in the vibration, rotation or state of the electron energy level of the molecule. If we want to kill cells or cause cell mutation, we must rely on ionizing radiation.

The attack of the dwarf corps on the organism, similar to other ionizing radiation, also needs to be achieved through direct or indirect interaction with various members of the cell factory. Here take the attacking process of DNA molecules by radiation as an example. Direct action means that the radiation directly breaks the chemical bond between the atoms or molecules on the DNA molecule. And the indirect action means that the radiation first interacts with hydrone in the environment to generate free radicals (OH·), which are then attached to the DNA molecules. The chemical groups interact with each other, which in turn causes the DNA strand to break. Some changes can be repaired by the "doctor of enzyme " and some cannot be repaired. Among so many members of the cell, why is DNA the most important targeted molecule? The metabolism of organisms is a dynamic process. After the action of ionizing radiation on sugar, protein and lipid molecules, the nature of these molecules get changed: some of them are recognized by the immune system of the organism, and others can be degraded by various enzymes. After a while, the changed molecules are wiped out in the living body. However, a DNA molecule carrying genetic information, if failing to repair properly after being attacked, will produce a change that is different from the previous one, i.e., the mutation. The expression of such genes is disturbed and the expression of proteins is affected accordingly. Therefore, the organism will exhibit a different phenotypic change than normal, some of which can be inherited through reproduction to the next generation. Then after several generations, new varieties will be born. Furthermore, as long as the time is long enough, new species will also be created (Figs. 4 and 5).

(4)   What Happens to the Contemporary Generation of Plants after They Are Irradiated?

Don't underestimate these invisible ionizing radiations. They are very lethal, especially the dwarf corps. If the plant is irradiated, growth will slow, stagnate or die and seeds may not be formed. Scientists have found that different plants have different levels of tolerance to radiation. Some are radiation-resistant, such as a kind of pasture called "alfalfa". Its dry seeds need more than 1000 Gy γ ray radiation to reduce the

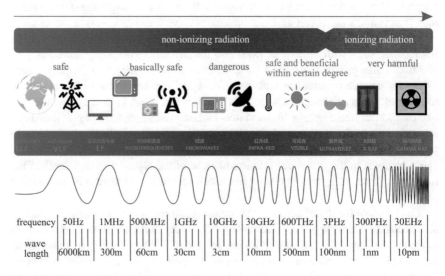

Fig. 4 Radiation of different types

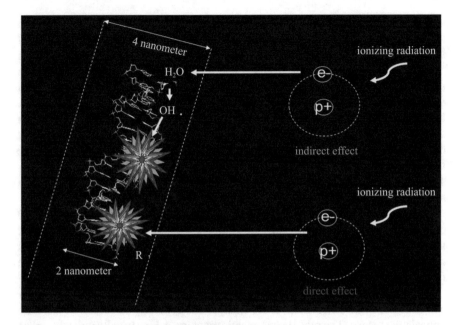

Fig. 5 Radiation and interactions among DNA molecules

growth height of seedlings by 50%; while some are not radiation tolerant, such as green onions, for whose seeds within 100 Gy γ ray radiation is enough. The difference between the two is 10 times. The degree of radiation sensitivity of plants is related to water content, DNA content, chromosome multiples, and different cell types. Plants are very radiation tolerant compared to animals.

Take an adult as an example. The probability of death is more than 50% if he receives 5 Gy of γ-ray radiation once a whole; and if he receives 10 Gy of γ-ray radiation once, the probability is close to 100%.

(5) Genetics and Variation

I go back to the library, brew a cup of coffee, open the multimedia computer, access the cloud database, and continue to search for documents. It is necessary to learn more about heredity and variation. The organism reproduces its offspring through various reproductive methods. Single-celled organisms breed themselves through cell division, and multicellular organisms reproduce through asexual reproduction and sexual reproduction. In either way of reproduction, life is guaranteed to be continuous generation after generation, and "children and grandchildren" are ensured to be similar to their "parents". This similar phenomenon between generations is "genetics". Genetics is a phenomenon in which genetic information is transmitted from generation to generation. The same species can only translate the same species of organisms, which is the so-called "you reap what you saw".

DNA is a very long molecule and is frequently damaged. Usually, there is an infirmary in the cell factory, and an "enzyme" doctor will be sent to repair it. But if the damage is extensive, it produces a permanent genetic code that can be called "mutation." If mutations occur in germ cells, they can be passed on from generation to generation. Mutation creates new features in biology. The replication and repair of DNA is important because the survival of organisms depends on a stable genome. However, not all errors can be repaired correctly. If the repair is wrong, the cells can still survive. This is the "variation". In fact, any organism in nature is mutating. But of course, the frequency is very low. When the environment changes drastically, the information contained in the original DNA blueprint cannot meet the needs of living, and some genetic variations can promote the creature to survive the storm and be naturally selected to survive. This is also the wonder of variation.

Certain chemical treatments increase the rate of mutations. These substances are called mutagens and the mutations they cause are called induced mutations. Most mutagens work directly by modifying a particular base of DNA or inserting into a nucleic acid. It is possible to introduce some mutations into any gene by applying a mutagen, but the specific mutations are random.

Variations can be divided into two categories. One is the changes in the number and structure of chromosomes. These changes can generally be perceived under an optical microscope. The other one is genetic mutation, which usually refers to a gene inside a specific site or small DNA fragment mutations that are not visible under microscope. But most of this kind can affect gene products and even alter the phenotype. Traditionally, the term "mutation" refers to genetic mutations, while the more obvious chromosomal changes are called chromosomal variations or chromosomal aberrations.

The variation of chromosome structure mainly includes the following four types. One is missing, meaning that the chromosome loses a certain fragment; the second is repetition, meaning that the chromosome increases a certain fragment; the third is inversion, meaning that the chromosome fragment is reversed 180 degrees, causing the rearrangement of genes in the chromosome; the fourth is translocation, meaning that non-homologous chromosomes exchange chromosome fragments, causing rearrangement of genes between chromosomes.

For the change in the number of chromosomes, we first introduce the haploid. Haploids contain only one genome, and all have only a single set of genes, usually denoted by n. If there are 2 sets of chromosomes, it is $2n$, and so on, there are $3n$, $4n$, etc. Plant polyploids are larger in volume than diploids, because as the number of chromosomes increases, the volume amplifies accordingly. Polyploids are important in production. Have you ever wondered why some watermelon bought in the supermarket have no seeds? For example, the diploid watermelon to be used is treated with colchicine at the seedling stage to obtain a tetraploid. Using the tetraploid as the female parent and the diploid as the male parent, you can get the triploid plant and the fruit of the knot, which is the seedless watermelon we often say.

Gene mutations are ubiquitous in the biological world, and the traits that appear after mutation do not show a corresponding relationship with environmental conditions. Mutations occur in natural conditions are called spontaneous mutations. And those induced by humans with physical or chemical factors consciously are called induced mutation. The phenotypic changes that occur after mutation vary. According to the effect of mutation on the phenotype, it can be divided into four types. The first is the morphological mutation, which mainly affects the morphological structure of the organism, resulting in changes in shape, size, color, etc.; the second is biochemical mutation, which can affect the biological metabolic process, leading to a specific biochemical function change or loss; the third is a lethal mutation, mainly affecting the vitality, leading to individual death; and the fourth one is the conditional lethal mutation, under which organisms can survive under certain conditions, and die under other certain conditions.

Mutations can occur at any time during the development of the individual organism and can result from spontaneous mutation or induced mutation. For example, in plants, buds are mutated in the early stages of development. After the buds grow into branches, leaves, flowers and fruits that are placed on top of them are different from other branches. This is the bud mutation. Bud mutation plays an important role in agricultural production, and many new varieties of fruit trees and flowers are made from bud mutations. Generally, the bud mutation is limited to a certain trait or some related traits, and many other traits are the same as the original cultivar. Therefore, these excellent traits can be retained in a short period of time by using vegetative propagation methods such as grafting and beading, and then appropriately selected and bred into a new variety. With the application of carbon dwarfs' radiation, combined with asexual cutting propagation, a new wonderful flower called "winter flower, summer grass" is created. The leaves of this flower are pink in winter and green in summer. Studies found that a pigment called anthocyanin which controlling the color of leaves, was controlled by temperature. In addition, with the same technique, a geranium flower mutant whose color changed from red to white is cultivated.

The history of radiation-induced mutagenesis dates back to 1927, when Muller used X-rays to treat Drosophila sperm, proving that X-rays can induce mutations and significantly increase the rate of mutation. During the same period, Lewis John Stadler used X-rays and γ rays to treat barley and corn seeds with similar results, which were published in the internationally renowned journals—*Science* and *PNAS* (Fig. 6).

2 Relevant Radiation Concepts and Radiation Facilities

Absorbed Dose

Different absorbed doses have different effects on creatures. The higher the doses are, the better the effect is. The absorbed dose relates to the energy of "radiation army" and properties of radiated substances.

Radiation Around Us

The environment we live in is full of background radiation. Nobody can avoid it. Our bodies have adapted well to natural background radiation for a long period. An adult (except for special populations) is exposed to average 2,400 microsievert

186 SCIENCE [Vol. LXVIII, No. 1756

SPECIAL ARTICLES

MUTATIONS IN BARLEY INDUCED BY X-RAYS AND RADIUM

At the Nashville meeting of the American Association last December I reported the occurrence of mutations in barley following X-ray treatment.[1] The experiments, which were independent of and coincident with those of Muller,[2] though by no means so comprehensive and thorough, confirm Muller's discovery of the power of X-rays to induce mutation and show its application to plants. They show also that mutations may be induced similarly by radium treatment.

The treatments were applied to germinating seeds, and the induced mutations recorded were in all cases somatic mutations affecting the progeny of only part of the plant. The experiments were set up in this way in order to exclude the possibility that new characters appearing after treatment might be ascribable to some irregular segregation from hybrid ancestry. The barley plant produces several tillers from axillary buds, each tiller terminating in an inflorescence of about thirty self-fertilizing flowers. In the dormant embryo the first three or four leaves are already differentiated, and the cells from which the tillers will be developed are separated. A mutation occurring in one of these cells, therefore, will affect only one tiller, and, whether dominant or recessive, will segregate in the progeny of only one head.[3] Its absence in the progeny of other heads of the same plant shows that the genetic change occurred during the develop-

were applied simultaneously, at target distances of 22.7 and 45.4 cm, respectively. The radiation passed through two samples of seed at shorter distances, and the filtering effect of the wet blotters and seeds must be considered in computing dosage. Ionization measurements made later showed that this reduced the intensity of the radiation at the higher voltage by about 52 per cent, and of that at the lower voltage by about 65 per cent. The relative ionizing intensity of the heavy and light treatments at the higher voltage and the heavy and light treatments at the lower voltage was in the ratio 100: 21: 50: 9. The so-called heavy doses were not heavy enough to reduce viability appreciably, but a dose of approximately three times this intensity, with the higher voltage, was found to be partially lethal. I am indebted to the department of physics of this university for the use of the X-ray equipment, and particularly to Mr. R. T. Dufford, of that department, for much advice and assistance and for the construction of an ionization chamber of the Duane type,[4] with which the dosage measurements above were made.

The radium treatments were applied under similar conditions, using as a source 50 mg of radium in the form of radium sulfate, sealed in a thin glass tube within a tube of silver 1 mm thick. Dr. Dudley A. Robnett, of Columbia University, generously lent his personal supply of radium for the treatments. The seeds, germinating in stacked watch glasses, were exposed continuously for twelve or twenty-four hours at distances ranging from one and a half to eleven cm. The maximum dose (applied to seeds in the

Fig. 6 Lewis John Stadler and its research result of Mutation of barley with X-rays published on *Science*

(μSv) a year of which over 50% originates from radon (RA), 15% originates from background radiation and 15% originates from medical natural background radiation. For example, the mean radiation dose to an adult from a chest X-ray inspection is around 20 μSv. Radon which we mainly absorb is radon-222. It is colorless and tasteless and it goes through alpha decay in our bodies. It is considered as a carcinogen (Fig. 7).

Epidemiology determines that absorbing radon in high concentrations is related to the incidence of lung cancer. Thus, radon is considered as a pollutant which impacts the global indoor air quality. According to the United States Environmental Protection Agency (EPA), radon raises the risk of lung cancer. 2,1000 people die of lung cancer in the United States every year. The food we eat contains radioisotopes with various radioactivity. For example, potassium-40 decays to argon 40 and calcium 40 giving off beta-rays and γ-rays. The radioactive dose of potassium-40 in every 500 grams of bananas is 65 Becquerel. Our bodies are also radioactive and contain radioisotopes with various abundances. The radioactivity of Carbon-14 and Potassium-40 in bodies of adults weighing 70 kilograms are 3,700 and 4,000 Becquerel respectively. Besides, there are polonium-210, radium-266, thorium, tritium, uranium, etc. in our bodies. We are also exposed to various levels of radiation on a flight. Scientists calculate theoretically that, for example, the total effective dose of an adult flying from Beijing to Chicago is 82 μSv.

Median Lethal Dose

The most common radiation dose used in mutation breeding is "lethal dose, 50%". The dose kills half of seeds of tested samples after radiation, which results in more mutant materials. These mutant materials have to change themselves to survive. So the mutation appears.

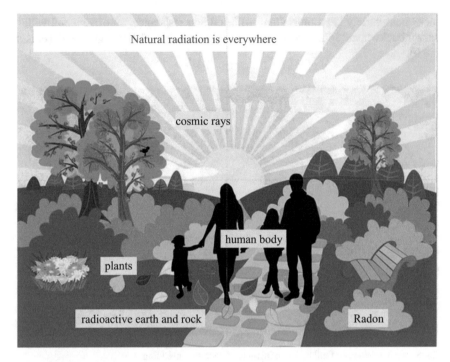

Fig. 7 Radiation around Us

Radiation Targets

The quality of most crops and industrial microorganisms can be improved after radiation. But radiation mutation does usually not apply to invasive plants and pathogenic microorganisms. Besides, some creatures' genetic manipulations are so complex that their mutant progenies tend to be unstable or they have already owned high spontaneous mutation rates without radiation.

Conventional Ray Radiation Devices

Common radiation devices are X-ray machines, γ-ray source devices, neutron source devices and heavy ion radiation facilities. Heavy ion radiation facilities can be mainly divided into two categories. The first kind of heavy ion radiation facilities are low energy ion implantation machines. They provide ion beams measured in electron volts which equals to the rifle's firepower. The second kind is medium and high-energy heavy ion accelerator facilities, the ion beams generated by which are measured in mega electron volts or billion electron volts. The latter kind has more power than the former and the energy is as powerful as the heavy artillery.

Now looking down from the flight, we could see Japan where the world's largest γ-ray irradiation breeding facilities locate. The γ-ray irradiation breeding field in the Institute of Radiation Breeding (IRB) is the world's biggest circular irradiation field with a radius of 100 meters which is longer than that of a football field. The γ-ray irradiation has demonstrated usefulness in releasing thousands of valuable mutations and large amount of new fruits and crops such as a new dwarf tartary buckwheat variety "Darumadattan" with lodging resistance, an ornamental yellow tartary buckwheat "Ionnokosai" and an ipomoea indica "IRBIi light blue" (Figs. 8 and 9).

Fig. 8 γ-ray irradiation field in Japan institute of radiation breeding

Fig. 9 RS-2400X irradiator device of RAD source technologies in USA

Fig. 10 McClellan nuclear research center of university of California

The γ-rays are generated by some special metals which are called radioactive sources, such as cobalt-60 and caesium-137. Radioactive sources including lead and concrete are very dangerous so that perfected radiation protections are needed. Because radioactive sources continuously decay, the only way to segregate them from people is shielding. Another kind of photon radiation, X-ray irradiation, is easier to operate than γ-ray irradiation (Fig. 10).

X-rays can be generated when hitting metal targets with X-ray tubes or electron accelerators. This process can be stopped at any time. The subsequent recycling process is easier for radioisotopes are not used in the whole process. There are many kinds of radiation mutation breeding facilities, among which RS-2400X irradiator device of RAD Source Technologies in USA is taken as an example.

McClellan Nuclear Research Center of University of California owns neutron irradiation facilities which can do researches on neutron activation analysis, radiation tolerance analysis, isotope production and fast neutron radiation breeding. Fast neutron irradiation plays an important role in plant mutation breeding and abundant mutant materials and new species of plants including arabidopsis thaliana, nicotiana tabacum, cucumber and soybean are harvested in the radiation process.

3   I Have Been Looking for Her for Thousands of Times in Vain

When I Turn My Head, Where Is She?

Plant seeds are called M1 generation (G-M1) before being radiated. After the dwarf-radiation process, it is called G-M1 seeds. We can harvest G-M1 plants if we sow G-M1 seeds in croplands. There are many physical damages on G-M1 plants, such as germination rate decline, survival rate decline, change of plant height, plants abnormality, etc. Regardless of those damages, breeders harvest G-M2 seeds from the G-M1 plants. They sow G-M2 seeds in croplands again and observe the changes of height, morphology, yield and harvest time. Then breeders select mutants from G-M2 plants and harvest G-M3 seeds from each mutant. They establish a mutant seeds databank by repeating this process, in which various needed materials can be found. After more genetic seed selection process for having more advanced-generations, breeders eventually obtain some new species which have high productivity. They hand out new species to farmers for satisfying the growing population's need for food, vegetables and fruits.

Please do not undervalue the selection process for ideal materials. Breeders cannot be easily found them until they observe thousands even millions of seeds. The most typical and common way of selection is depending on breeders' experience without any equipment, which makes selection work highly intensive. As science and technology develop, selection work can partly be done by machine nowadays. The age of big data which is characterized by high-throughput, no-damage, automaticity and high-precision brings another revolution to plant breeding history. In just a few years, dozens of companies and research institutes including Dupont, Monsanto, Donald Danforth Plant Science Center, Leibniz Institute of Plant Genetics and Crop Plant Research of America (IPK), European Plant Phenotyping Platform (PhenoFab), Institute of Physical and Chemical Research of Japan (RIKEN), Bavarian State Research Center for Agriculture of Germany (LFL), LemnaTec of Germany, have researched and developed characteristic high-throughput phenotyping platforms, which greatly reduce work intensity.

With the coming of the era of artificial intelligence, various robots will emerge in farmland in the near future. Robots can observe and record changes of plant mutations on the hot summer day while breeders in the future will analyze data collected by robots in a cool air-conditioning room. Those well-equipped robots with many detective cameras can detect the visible light, the infrared light and fluorescence then they depict three-dimensional images. Robots scan phenotype of hundreds of plants rapidly within a very short period of time. They seem to say to mutations that "Once you change, I will find you immediately" (Fig. 11).

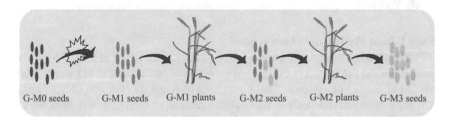

G-M0 seeds   G-M1 seeds   G-M1 plants   G-M2 seeds   G-M2 plants   G-M3 seeds

Fig. 11 Flowchart of radiation mutation breeding

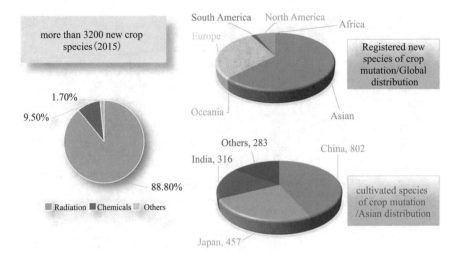

Fig. 12 Mutation breeding results and its distribution

4 Mutation Breeding Shows Power

Plants

According to the International Atomic Energy Agency (IAEA) and Food and Agriculture Organization of the United Nations (FAO), more than 3,200 new crop species are harvested by mutation breeding, 88.8% of which are mutated by radiation. Asian countries prefer improving plants species in the way of mutation breeding. For instance, countries like China, Japan and India have released hundreds of new species with high-quality (Fig. 12).

5 New Weapons of Radiation Mutation Breeding

Heavy Ion Beam Mutation

The army formed by thousands of dwarfs is called heavy ion beam. It is a kind of ionizing radiation. Compared with photon irradiation such as X-ray irradiation and γ-ray irradiation, ionizing radiation releases more energies and leads to more complex damages to organisms. The principle of dwarf breeding can be elaborated in following process. Carbon dwarfs, for example, each nucleus of which contains 80 mega-electron volt (MeV), can radiate plant seeds.

These dwarfs pass through cells successively waving their broadswords (with energy). When they come into the cell nucleus, they interact with atoms of DNA molecules and cut chemical bonds directly or they just interact with molecules of water in cells and generate new soldiers—free radicals. Old and new dwarf soldiers cut off the blueprint of cells—chromosomes, so various forms of breaks in different

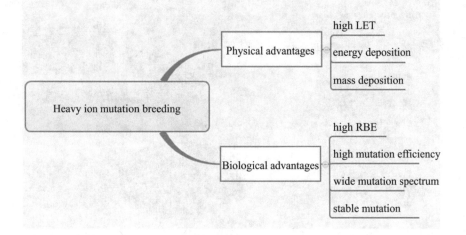

Fig. 13 Traits of dwarf army mutation breeding

parts emerge. Particular doctor Enzymes inside the plant cells come fast to repair breaks. Some breaks are easy to repair. They will be fine once they are connected with each other again. For those who cannot be cured, many doctor enzymes have to make a consultation in limited time. They resort to related information or repair with previous experience. Then wrong reparations happen. Sometimes even one small mistake causes a disaster, and the plant will not go back to its original form. Thus the mutant cells appear and form their own community, in which they help each other and grow together to an integrated plant (the new mutant). Mutants may be taller or shorter (the height mutation), be precocious or late-maturing (maturation stage mutation) or just breed more (yield). They are selected in laboratories and in the farmland and better mutants will be picked out to become new species (Fig. 13).

Dealing with biological samples by using dwarfs introduces more double-strand breaks (DSBs) and clustered damages in DNA. Those damages are much harder to repair than those which are caused by the same dose of photon irradiation (like X-rays and γ rays). The wrong reparations and recombination generate various types of changes in chromosome structures. They set closed linked genes apart, which leads to gene mutation and obtains new mutation types. In short, the dwarf army radiation mutation breeding has higher mutation efficiency, wider mutation spectrum and shorter stable period.

Global Heavy Ion Radiation Breeding Facilities

There are four accelerators generating the dwarf army in Japan, which are used to do researches on radiation mutation breeding. They are Nishina Accelerator System in RIKEN, Heavy Ion Medical Accelerator in Chiba (HIMAC) in National Institute of Radiological Sciences (NIRS) of National Institutes for Quantum and Radiological Science and Technology (QST), Takasaki Ion Accelerator for Advanced Radiation Application (TIARA) in Takasaki Advanced Radiation Research Institute of QST, Wakasa Multi-purpose Accelerator with Synchrotron and Tandem (W-MAST) in the

Wakasa Wan Energy Research Center (WERC). Japan has established a specialized association named the Ion Beam Breeding Society (IBBS) to push the application of the technology of ion beam breeding on its fundamental and applied researches. As a result, it produces a large amount of new plant species.

Heavy Ion Research Facility at Lanzhou (HIRFL) developed by the Institute of Modern Physics (IMP) of Chinese Academy of Sciences (CAS) is the only big science facility in China. It has capability to provide the medium energy and high energy heavy ion beams. Staffs from dozens of scientific research institutions come to IMP every year to do mutation breeding fundamental and applied researches and make a large amount of achievement. The HIRFL contains two cyclotrons and two synchrotrons. Cyclotrons include Sector Focusing Cyclotrons (SFC) and Separated Sector Cyclotrons (SSC). Cyclotrons accelerate carbon dwarfs (C) and uranium dwarfs (U) up to 10 MeV/u and 1.08 MeV/u respectively through electromagnetic fields. SSC does a further effort to motivate dwarfs run faster and it accelerates carbon dwarfs (C) and bismuth dwarfs (Bi) up to 100 MeV/u and 9.5 MeV/u separately through electromagnetic fields. Heavy ion beam breeding always chooses the carbon dwarfs which are 80 MeV/u to obtain high-quality mutation materials or new species through radiating various plants materials such as seeds, branches, leaves, roots, tubers, tissues, suspension cells, etc. (Fig. 14).

Unexpected Progeny—Plant Mutation Breeding

Dry crop seeds are usually chosen in radiation process for they are easy to handle. Different types of crops have different radiation sensitivities to dwarfs. Some crops

Fig. 14 Plant seeds radiated by dwarfs

vary with only little radiation doses while others need dozens or hundreds times of doses. The younger the plant tissues are, the more sensitive to radiation they are. So are those materials containing more water. Thus, wet seeds and tender shoots are sometimes selected to under radiation. In order to generate more DNA mutations, researchers explore different types of crops for suitable radiation doses and biological samples. Certainly, to clarify the principles of how the dwarf army effects the plant growth and mutation, scientists need to do systematic studies on simplest plants which have short growth cycles.

(1)   Model Plants

Model plants are a species of plants regarded as the symbol of all plants. People know plants secrets of birth, death, illness and death by exploring model plants. People take this species of plants as a model to deduce the law of life activities based on the analysis and conclusion of morphologies, anatomical structures, physiological functions, chemical processes, cell states and genetic mechanism of modal plants. Common model plants include arabidopsis thaliana, nicotiana tobacco, lotus japonicus, antirrhinum majus, petunia, oryza sativa, corn, etc. When it comes to one typical representative of model plants, arabidopsis thaliana cannot be ignored. Arabidopsis thaliana is in the mustard family, within the same family there are cabbage and rapeseed. Compared with them, arabidopsis thaliana has a shorter height and its seeds are as small as pinpoints. So, why the arabidopsis thaliana can be chosen as the typical model plant among numerous plants species? It has several traits that make it a useful model plant.

①   The plant's small size and rapid lifecycle are advantageous for research. The height of arabidopsis thaliana is short ranging from 30 to 40 cm. An arabidopsis thaliana can complete its entire lifecycle, from sprout, growth, blossom to death, within 8 weeks. It annually breeds seeds for 6 times. If scientists do research on a tree, it will take dozens of years from planting saplings to mature the first generation seeds.

②   The plant has large seeds production. Although it has a small size and a short lifecycle, each plant produces around 5000 seeds, which seems make it "vulgar" in nature.

③   The small size of its genome with abundant genetic information makes arabidopsis thaliana useful for research. The work of whole genome sequencing on arabidopsis thaliana had been done in 2000—with five chromosomes and 125 mega base pairs. The plant has one of the smallest genomes among known higher plants. Genomes of each creature can be seen as a handbook which introduce the way to make the creature. Most of them contain enormous information, but most information is meaningless. The useless words cover some crucial information. However, the handbook of arabidopsis thaliana is much thicker and contains less meaningless words (Figs. 15 and 16).

(2)   How the Dwarf Army Changes Model Plants?

(*Arabidopsis thaliana*) (*Lotus corniculatus*) (*Nicotiana tabacum*) (*Antirrhinum majus*)   (*Oryza sativa*)

Fig. 15 Various types of model plants

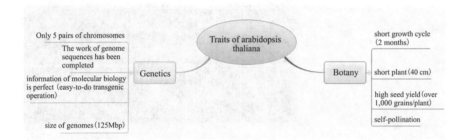

Fig. 16 Model plant arabidopsis thaliana

Since model plants are typical, scientists think the model plants' reaction to physical radiation can representative the basic reaction to radiation breeding, and they thereby can deduce the basic principle and mechanism of plant radiation mutation. What expressions do model plants have under the radiation? In room temperature and atmospheric conditions, we use carbon dwarfs (43.3 MeV/u, 200 Gary dose) generated by HIRFL to irradiate arabidopsis thaliana seeds whose following generations have different types of mutation. To sum up, the mutations fall into four categories: leaf mutation, stem mutation, flower mutation and lifecycle mutation. The whole mutation rate of carbon dwarf irradiation is up to 4.77% while the rate of γ-ray irradiation is only 1.92%. The radiation mutation rate of the carbon dwarf doubles that of γ-rays. Also, carbon dwarfs can induce more types of mutation such as flower mutation and stem mutation (Fig. 17).

(3)   What are the differences between mutants and the original form?

Modern biological technology is becoming increasingly advanced. The genome sequencing skill can read the whole sequence of DNA blueprint. The root cause of mutation lies in explaining varied arabidopsis thaliana seeds at the genome level by adopting the high-throughput sequence skills. We find the genomes of plant mutants, which is referred as the handbook of creatures, change in many parts. The mutation is different from "the plastic surgery" which only changes the human's appearance

Fig. 17 Arabidopsis
thaliana mutants irradiated
by carbon dwarfs

in human community. It superficially changes the plant morphology and traits, but actually it is a change of genomes or genetic information under radiation. It not only changes the outside but the inside. So there is no need to worry about the durability of mutation.

Crops

Because carbon dwarfs generate more DNA mutations and the rate of which is high, so they are widely used in crop breeding. Seeds after radiation show a series of changes in growth for the DNA mutations cause a series of changes in the morphology, the yield, the quality and the resistance. Breeders select the helpful mutants to cultivate new species to fulfill the different requirements through field observation and instrumental analysis (Fig. 18).

Researchers use carbon dwarfs to irradiate crops to harvest abundant mutants and cultivate new species which are featured by high-yield, high-quality and resistance, such as oryza sativa, wheat, sorghum, traditional Chinese medicinal materials, etc. Planting a large scale of these new species brings enormous economic benefits (Figs. 19, 20, 21, 22, 23, 24, 25 and 26).

Ornamental Plants

Ornamental plants are plants with ornamental value which are grown for decorating, landscaping and improving environment as well as enriching people's life. With the improvement of living standard, people's pursuit of novel flowers is surging. Flowers production and application have continuously developed in recent years.

Fig. 18 Sorgo—new precocial species

Fig. 19 Sunflower—flower shape mutation

Fig. 20 Wheat—Spike form mutation

Fig. 21 Soybean—Grain
mutation

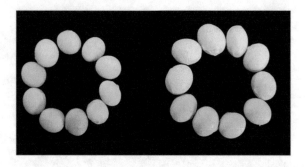

Fig. 22 Corn—Increased
ear row number

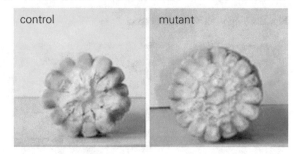

Fig. 23 Oat/Avena
sativa—Spike form mutation

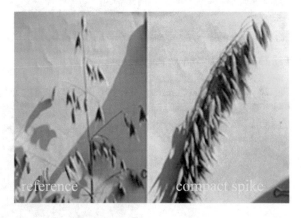

Independent intellectual property rights for flowers can be highly protected when we get some new species by using artificial mutagenesis.

Japanese researchers firstly use the dwarf army to radiate flowers and get abundant mutants. There are lots of successful examples. For instance, nitrogen dwarfs are used to irradiate verbena. Then there appears a new commercialized species which has longer blooming period, more flowers and sterility which brings enormous economic benefits. Red, pink or dichromatic mutants of carnation leaves are selected after

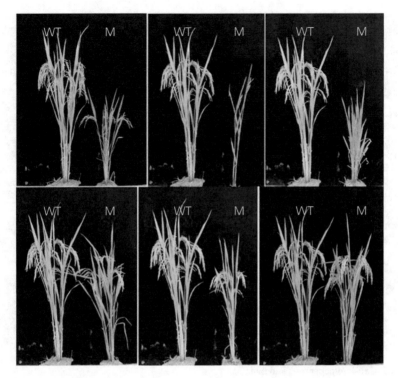

Fig. 24 Oryza sativa—Plant type mutation

being exposed to carbon dwarfs' radiation. They have been registered by Japanese Agriculture Ministry and are allowed to launch on the market. These also bring huge economic benefits. Otherwise, some plants alter to striped colored, complex colored, multi-petalled, and petal-decreased mutations after their petals and stems are radiated by dwarfs. Chrysanthemum, for example, has some complex colored, multi-petalled, axillary bud decreased mutants which can bloom in low-temperature environment and some new germplasm resources. Institute of Modern Physics of Chinese Academy of Sciences adopted advanced heavy ion beam mutation technology and had obtained a new species named "winter flower, summer grass" of tradescantia fluminensis whose leaves color changes with seasons and a species of geranium mutation which has an apparent color change.

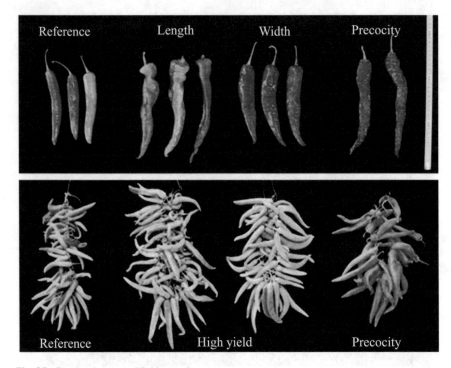

Fig. 25 Capsicm/pepper—Yield mutation

Fig. 26 Ricinus communis—morphologic mutation

The new species tradescantia fluminensis is a color-leafed plant whose leaves change with temperature. Its stems are purple and leaves have pink spots in winter while they turn to green in spring and summer. That is why people call it "winter flower, summer grass" (Figs. 27, 28 and 29).

Others

Algae is a lower autotrophic plant which is characterized by a wide distribution both on land and in the sea, the rich nutrition, and high photosynthetic efficiency.

Fig. 27 Hrysanthemum mutation irradiated by carbon dwarfs

Fig. 28 "Winter flower, summer grass"; leaves turn pink in winter (A2, B2); leaves turn green in spring and summer (A3, B3)

Fig. 29 Geranium mutation irradiated by ion beam

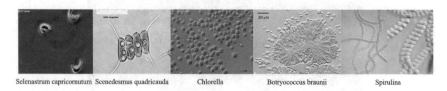

Selenastrum capricornutum Scenedesmus quadricauda   Chlorella   Botryococcus braunii   Spirulina

Fig. 30 Microalgae mutation irradiated by carbon ion beam

Unlike other plants which have stems and leaves, it is very simple plant without those tissues. It owns a thallus, an organelle which is similar to chloroplast, to do photosysthesis. Algae provides many proteins, lipid, algae polysaccharides, beta carotene and other inorganic elements. It can be made into not only edible nutriments but chemical materials. Microalgae is applied widely in the field of biological energy, environmental manipulation, food and medicine, etc. Existed researches prove that algae mutations radiated by carbon dwarfs contain over 30% more oil than the original form, have higher growth rate and higher photosynthetic efficiency than before. At present, the common kind of microalgae used in carbon dwarfs breeding is unicellular green algae. Japanese researchers select and breed striped laver mutation with pigment changes. The microalgae mutations which are selected by the Institute of Modern Physics of Chinese Academy of Sciences through heavy ion beam irritation are selenastrum capricornutum, scenedesmus quadricauda, chlorella, botryococcus braunii as well as spirulina. Algae mumations with higher oil content and higher growth rate of have been found (Fig. 30).

6 Safety Inspectors of Spaceship

When you travel in the space on a spaceship, have you thought about the attack of dwarfs, tiny dwarfs and γ-ray from many remote galaxies to you in every second? Also, each part of integrated circuit on the computer which controls the space forward are attacked by the dwarfs and rays in every moment. This is the price you should pay for travelling in the space. Even though, people are willing to make effort to seek secrets of exoplanets such as Mars and Sirius and try to find which planets are suitable for human beings to live.

Dwarfs wandering the vast space of solar system are proton dwarfs, helium dwarfs, other tiny dwarfs like neutron and neutrino and a few relative heavy dwarfs from the sun. Their speeds are fast, almost equal to the speed of light. So, there is nothing can stop them. When they pass through space travelers' bodies, faster speeds make less damage. However, a proton dwarf, a helium dwarf or a heavy dwarf with extreme speeds will hit the key part of spaceship control system and unfortunately make electrons in this part out, which is called Single Event Effect (SEE) by scientists. This will damage the quality or burn the key electronic components in hit part. Whatever happens, consequences lie in the abnormal flight or the route deviation without controls. How horrible the consequences are! Once the imagination comes to space travelers, they will shudder.

How can we prevent or avoid the accidents? First of all, every component should be tested through a specific examination to see whether they fulfill the requirements of entering the space. Then, they are further armed to resist the attack from exotic dwarfs.

Dwarf army coming from the SSC training center heads for the aims under the guidance of magnetic field. They go through the silicon castles, hit castles and knock down the electrons on the exterior walls of castles. As a result, their speeds of them slow down. Finally, they arrive in the electronic components market which is arranged intensively and orderly. But they ignore those electronic components and only stare at silicon castles. Each soldier goes through the market, strike castles and try to knock down the electrons on exterior walls of castles (Fig. 31).

Suddenly, a dwarf soldier pulls all the tricks to go through walls of castles successively in a castle complex near a current control valve and knocks down a great many of electrons. These electrons form an electron flow which has great powers turning on the valve. Plenty of electrons enter the pipe rapidly and burn as fires to damage other castles in the valve, which makes walls collapsed successively.

Several similar phenomena appear during the one-hour's dwarf army attack. The valve cannot work under attacks. However, the same accidents happen every twenty years in space. As a result, satellites which are designed for ten-year operation can be used (Fig. 32).

In recent ten years, researchers of IMP of CAS have cooperated with China Aerospace Science and Technology Corporation, China Light and Power Company (CLP), China Academy of Science and universities. They used fat high-energy dwarfs trained by dwarf army sequence training center (heavy ion accelerator) at Lanzhou

Fig. 31 Space radiation environment

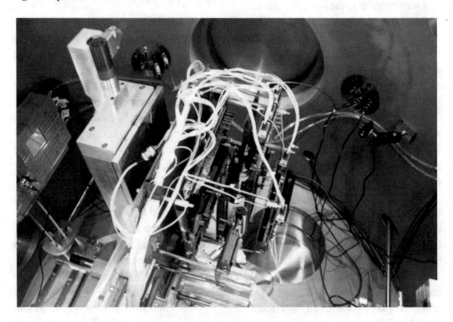

Fig. 32 Single Event Effect experiment with Heavy ion accelerator at Lanzhou

to inspect many kinds of integrated circuit components in the field of spaceflight and to assess the risk of space flight. They collected a mass of data which have great value to the study of aerospace components and system radiation hardening.

Events About Heavy Ion Research Facility in Lanzhou (HIRFL)

1950	Institute of Modern Physics (IMP) of Chinese Academy of Sciences (CAS) was founded in Beijing, currently known as Institute of Physics
1956	Premier Zhou Enlai gave instructions that a nuclear science and technology institiue should be founded in Lanzhou. Then Mr. Yang Chengzhong, a famous nuclear physicist, prepared to build Lanzhou Physics Laboratory in Beijing
1957	The Lanzhou Institute of Modern Physics of Chinese Academy of Sciences was founded in Lanzhou
1958	The the Second Ministry of Machine (currently known as Ministry of Machinery Industry of the People's Republic of China) founded the 613 Project Office of Chinese Academy of Sciences (known as 613 Project), which was responsible for the construction of 1.5 m cyclotron, one of 156 important industrial projects with the aid of the Soviet Union in China's First Five-Year Plan
1962	Lanzhou Physics Study Office of Chinese Academy of Sciences amalgamated with the 613 Project Office of Chinese Academy of Sciences, forming the Institute of Modern Physics, Chinese Academy of Sciences, the code of which was "Northwest 203 Office"
1976	State Planning Commission formally approved to establish a large-scale separated sector heavy ion accelerator by IMP of CAS in the Seventh Five-Year Plan, the code of which was "7611 Project"
1989	The 7611 project passed the national technological appraisal and evaluation at its completion
1991	State Planning Commission approved to establish Heavy Ion Research Facility in Lanzhou (HIRFL) National Laboratory. The laboratory was opening and sharing and have provided experiment conditions for over one hundred domestic and international users
1997	China's first Radioactive Ion Beam Line at Lanzhou (RIBLL) was established
1998	State Planning Commission formally approved to establish one of the national great science engineering projects, the Cooler-Storage-Ring at Heavy Ion Research Facility in Lanzhou (HIRFL-CSR)
2008	HIRFL-CSR passed the national technological appraisal and evaluation at its completion

© Zhejiang Education Publishing House 2021
G. Jin and G. Xiao (eds.), *Probing into the World of Nuclei*, China's Big
Science Facilities, https://doi.org/10.1007/978-981-16-0715-8

So far, HIRFL has become China's largest scale facilities which contain the most diverse accelerated ions and the highest energy. The main technological index has gone up to an international advanced level. HIRFL consists of the Sector Focusing Cyclotron (SFC), the Separated Sector Cyclotron (SSC), the Cooler Storage Ring (CSR) and a number of experimental terminals. It has great power to accelerate complete ions and provides various stable ion beams and radiant beams which have wide energy range, and high quality. HIRFL provides advanced experimental conditions for national heavy ion physics study and its inter-disciplines.

Printed in the United States
by Baker & Taylor Publisher Services